The Respiratory System

be well
informed

@ your library®

ALA American Library Association   Walgreens The Pharmacy America Trusts

This book was received through a grant from
Walgreens and the American Library Association.

**Other titles in
Human Body Systems**

The Respiratory System

David Petechuk

HUMAN BODY SYSTEMS
Michael Windelspecht, Series Editor

Greenwood Press
Westport, Connecticut • London

Library of Congress Cataloging-in-Publication Data

Petechuk, David.
    The respiratory system / David Petechuk.
        p.   cm.—(Human body systems)
    Includes bibliographical references and index.
        ISBN 0–313–32434–4 (alk. paper)
        1. Respiratory organs—Diseases.   2. Respiratory organs.   3. Respiration.   I. Title.
    II. Human body systems.
    RC731.P485   2004
    612.2—dc22   2004040445

British Library Cataloguing in Publication Data is available.

Library of Congress Catalog Card Number: 2004040445
ISBN: 0–313–32434–4

First published in 2004

Greenwood Press, 88 Post Road West, Westport, CT 06881
An imprint of Greenwood Publishing Group, Inc.
www.greenwood.com

Printed in the United States of America

The paper used in this book complies with the
Permanent Paper Standard issued by the National
Information Standards Organization (Z39.48–1984).

10 9 8 7 6 5 4 3 2

Illustrations, unless otherwise credited, are by Sandy Windelspecht.

The *Human Body Systems* series is a reference, not a medical or diagnostic manual. No portion of this series is intended to supplement or substitute medical attention and advice. Readers are advised to consult a physician before making decisions related to their diagnosis or treatment.

To my mother and aunt, Kathleen and Marjorie Yorko

Contents

Color photos follow p. 70.

Series Foreword

Human Body Systems is a ten-volume series that explores the physiology, history, and diseases of the major organ systems of humans. An organ system is defined as a group of organs that physiologically function together to conduct an activity for the body. In this series we identify ten major functions. These are listed in Table F.1, along with the name of the organ system responsible for the activity. It is sometimes difficult to specifically define an organ system, because many of our organs have dual functions. For example, the liver interacts with both circulatory and digestive systems, the hypothalamus acts as a junction between the nervous and endocrine systems, and the pancreas has both digestive and endocrine secretions. This complex interaction of organs and tissues in the human body is still not completely understood.

This series is unique in that it provides a one-stop reference source for anyone with an interest in the human body. Whereas other references frequently cover one aspect of human biology, from anatomy and physiology to the prevention of diseases, this series takes a more holistic approach. Each volume not only includes a physiological description of how the system works from the cellular level upward, but also a historical summary of how research on the system has changed since the time of the ancients. This is an important aspect of the series, and one that is frequently overlooked in modern textbooks. In order to understand the successes and problems of modern medicine, it is first important to recognize not only the achievements of the past but also the misunderstandings and challenges of the pioneers in medical research.

For example, a visit to any major educational institution reveals large lec-

TABLE F.1. Organ Systems of the Human Body

Organ System	General Function	Examples
Circulatory	Movement of chemicals through the body	Heart
Digestive	Supply of nutrients to the body	Stomach, small intestine
Endocrine	Maintenance of internal environmental conditions	Thyroid
Lymphatic	Immune system, transport, return of fluids	Spleen
Muscular	Movement	Cardiac muscle, skeletal muscle
Nervous	Processing of incoming stimuli and coordination of activity	Brain, spinal cord
Reproductive	Production of offspring	Testes, ovaries
Respiratory	Gas exchange	Lungs
Skeletal	Support, storage of nutrients	Bones, ligaments
Urinary	Removal of waste products	Bladder, kidneys

ture halls, where science instructors present material to the students on the anatomy and physiology of the human body. Sometimes these classes include laboratory sessions, but in the study of human biology, especially for students who are not bound for professional schools in medicine, the student's exposure to human biology typically centers on a two-dimensional graphic. Most educators accept this process as a necessary evil of the educational system, but few recognize that, in fact, the large lecture classroom is the product of a change in Egyptian religious beliefs before the start of the current era. During the decline of the Egyptian empires and the simultaneous rise of the ancient Greek culture, the Egyptian religious organizations began to forbid the dissection of the human body. This had a two-fold influence on medicine. First, the ending of human dissections meant that medical professionals required lectures from educators instead of participation in laboratory-based education, which led to the birth of the lecture hall. The second consequence would plague modern medicine for a thousand years. Stripped of their access to human cadavers, researchers studied other "lesser" animals and extrapolated their findings to humans. The practices of the ancient Greeks were passed on over the ages and became the basis for the study of modern medicine. These traditions continue to this day throughout the educational institutions of the world.

The history of human biology parallels the development of modern science. In the seventeenth century, William Harvey's study of blood circulation challenged the long-standing belief of the ancient Greeks that blood was produced in the liver and consumed in the tissues of the body. Harvey's pioneering experimental work had a strong influence on others, and within a century the legacy of the ancient Greeks had collapsed. In the eighteenth century a group of chemists who focused on the chemical reactions of the human body, called the iatrochemists, began to apply chemical laws to human physiology. They were joined by the iatrophysicists, who believed that the human body must operate under the physical laws of the universe. This in turn led to the beginnings of organic chemistry and biochemistry in the nineteenth century, as scientists focused on identifying the building blocks of living cells and the chemical reactions that they utilize in their metabolism.

In the past century, especially in the last three decades, the rapid advances in technology and scientific discovery have tended to separate most sciences from the general public. Yet despite an ongoing trend to leave the majority of the physical sciences to the scientists, interest in human biology has actually increased among the general population. This is primarily due to medical discoveries that increase not only lifespan but also healthspan, or the number of years that people live disease free. But another important aspect of this trend is the desire among the general public to be able to ask intelligent questions of their physicians and seek additional information on prescribed medications or procedures. In many cases, this information serves as a system of checks and balances on the medical profession, ensuring that the patient is kept well informed and aware of the fundamentals regarding the procedure.

This is one of the most remarkable ages in the study of human biology. The recently announced completion of the Human Genome Project is an indication of how far biology has progressed. Barely fifty years ago, scientists were first discovering the structure of DNA. They now are in possession of an entire encyclopedia of human genetic information, and although they are not yet exactly sure what the content reveals, scarcely a week goes by without a researcher announcing a medical discovery that was made possible by the availability of the complete human genetic sequence. Coupled to this are the advances in the development of pharmaceuticals and treatments that were unheard of less than a decade ago.

But these benefits to society do not come without a cost. The terms stem cells, cloning, and gene therapy no longer belong to the realm of science fiction. They represent advances in the sciences that may hold the key to increased longevity. However, in many cases they also produce ethical and moral questions of society: Where do medical researchers obtain the embryonic stem cells for their work? Who will determine if humans can be cloned? What are the risks of transgenic organisms produced by gene therapy? These are just a few of the potential conflicts that face modern society. Only a well-educated general public can intelligently survey the pros and cons of an ethical or moral decision regarding medical science. Armed with information, concerned people can participate in the democratic process of informing their elected officials of their concerns. Science education is an important aspect of citizenship, and thus the need for series such as this to present information to the general public.

This volume covers the biology of the respiratory system. There are actually two forms of respiratory systems in the human body. The familiar action of breathing involves the action of the lungs, diaphragm, and ribs. It is this system that is involved in the exchange of gases, including the life-giving oxygen. The second form of respiration is at the cellular level, and involves the use of oxygen to break down organic nutrients for energy. More than 63 trillion cells in the human body all require oxygen for their metabolic activities. In our modern world, our respiratory systems, both at the organ and cellular level, are under constant attack from pollutants in the air. Secondhand smoke and ground level ozone are constantly battering our respiratory systems, causing an increase risk of diseases such as cancer. Thus there is a definite need for an understanding of how the respiratory system fits into the overall workings of the human body.

The ten volumes of *Human Body Systems* are written by professional authors who specialize in the presentation of complex scientific ideas to the general public. Although any book on the human body must include the terminology and jargon of the profession, the authors of this series keep it to a minimum and strive to explain the concepts clearly and concisely. The series is ideal for the public libraries, as well as for secondary school and introductory college libraries. In addition, medical professionals or anyone with an interest in human biology would find this series a useful addition to their personal library.

Michael Windelspecht
Blowing Rock, North Carolina

Introduction

When the aging vaudeville entertainer Sophie Tucker was asked what the key to long life was, she replied, "Keep breathing." Although Tucker was making a joke, her answer was also correct in the most fundamental biological way. Although all of the human body's various systems are integral to life, none of them—from the cardiovascular to the nervous systems—would be able to function without the respiratory system. It is the respiratory system that garners the body's most basic fuel in the form of oxygen that we breathe in from the air. Every cell in our body uses oxygen to produce energy from food and drink. In fact, every chemical process throughout the body ultimately needs oxygen to take place. It is also through the respiratory system that the body eliminates carbon dioxide waste from cell metabolism. If the respiratory system ceases to function, death occurs within minutes as carbon dioxide rapidly reaches toxic levels in the blood.

When most people think of the respiratory system, they generally think of the relatively simple concept of breathing in and out, which is called respiration. But the respiratory system is a complex assemblage of organs and tissues that are integral to three different types of respiration. Breathing begins with nerve impulses that stimulate the breathing process, moving air into and out of the lungs through a series of passages from the nose down through the throat and into the lungs. Once the oxygen-rich air reaches the lungs, gas exchange (oxygen for carbon dioxide) occurs between the lungs and the blood. This process is called *external respiration*. Then, working in concert with the circulatory system, the now oxygen-rich blood is transported to all of the body's tissues where the gas exchange process occurs once again, this time between the blood and cells, with the blood passing oxygen into the cells and carrying away carbon dioxide to be eliminated via the lungs and expiration. This respiratory process is called *internal respiration*. Once the oxygen reaches the cells, it is used for a variety of specific energy-producing activities within the cells. This third form of respiration is called *cellular respiration.*

Our in-depth understanding of the respiratory system's various functions is relatively recent. Although our ancient ancestors knew the importance of breathing to life, they had little knowledge of the respiratory system itself or how it functioned. Most early beliefs about the respiratory system were based on philosophy, religious beliefs, and superstition. For example, both Plato and Aristotle believed that the heart contained a fire and that breathing kept this fire under control. Knowledge about the respiratory system early on was hindered by the fact that anatomical studies focused on animals and not humans, because anatomical study of the human body was taboo in most ancient cultures. Not until the seventeenth century would significant progress take place in understanding the very basic functioning of the respiratory system; another century would pass before oxygen was identified as the vital component in the respiratory process. Only during the twentieth century, with the advent of modern science, would scientists begin to understand the many intricacies of gas transport through the blood and cellular respiration.

The respiratory system, especially the lungs, is unique from other systems in that it is in close and constant contact with the outside environment via the air we breathe. As a result, it is exposed to a wide variety of potentially harmful substances, from naturally occurring bacteria and viruses to pollutants produced by humans and modern society. Largely because of these exposures, respiratory diseases and illnesses—from the common cold to asthma to lung cancer—are among the most prevalent forms of sickness and disease in human beings. In the United States, respiratory-related problems are the most common reason that people go to see their doctors. Respiratory illnesses as a whole are the third leading cause of death in the United States, and lung cancer is the country's leading cancer killer. Respiratory disease can also be especially difficult to treat in children and the elderly, due to smaller lungs and decreased lung functioning, respectively. For example, influenza ("the flu") and pneumonia are major causes of illness and death in older people. Taking into account allergies, colds, and the flu, few people ever go through life without experiencing some respiratory difficulties.

Ancient remedies for respiratory problems focused primarily on the use of herbs. Not until the 1940s did scientists discover antibiotics for effective treatment of common respiratory infections. Mechanical ventilators were also developed in the twentieth century to help people breathe who were suffering from severe respiratory problems. These machines were first widely used in response to the polio epidemic of the 1940s and 1950s.

Today, scientists are focusing on developing treatments based on the molecular physiology of respiratory diseases, including the influence that genetics has on causing diseases or making people more susceptible to certain respiratory ailments. For example, although cigarette smoking is the overwhelming cause of lung cancer, many smokers do not get lung cancer. (It should be noted, however, that many other deadly diseases are associated with smoking, such as emphysema.) Current research is delving into what role certain genes may play in placing some smokers at a higher risk of contracting this deadly disease.

Despite the advances made in treating respiratory diseases, they continue to be a major health threat. For example, an estimated 14.9 million persons in the United States have asthma, and the number of people with asthma increased by 102 percent between 1979 and 1994. While many respiratory ailments such as asthma result from factors beyond an individual's control, smoking cigarettes and other tobacco products represent a self-inflicted and deadly assault on the respiratory system, as well as other parts of the body. Smoking kills more people in the United States each year than car accidents, alcohol, AIDS, murders, illegal drugs, and suicide combined. Even though science has proven that smoking can be deadly, 23.5 percent of the United States population still smokes. In addition, smokers are also harming the people around them. A team of scientists from the World Health Organization (WHO) has estimated that exposure to secondhand smoke in nonsmokers increases their risk of getting lung cancer by 20 percent.

This volume provides an in-depth look at the respiratory system, from the basic anatomy and functioning of the system to related diseases and treatments. It also provides an historical perspective on early beliefs about breathing and the respiratory system and on the progression of scientific inquiry that led to our current understanding of how the system works. The final chapter focuses on two aspects of modern life that greatly affect the respiratory system: smoking and air pollution. At the end of this volume, a list of acronyms, a glossary, a list of organizations and Web sites, and a bibliography are provided. Terms included in the glossary are **bold** on first mention in the text.

Although this volume deals with some complex scientific information, it is written for a general audience—from high school and undergraduate college students to the adult reader who is simply interested in learning more about the many fascinating aspects of the respiratory system. Today, medical and biological science is advancing more rapidly than at any other time in history. Current research into the medical and biological aspects of the system is incredibly wide ranging, from molecular studies of proteins to the promise of altering a person's genetic makeup to help cure or prevent respiratory diseases. As a result, in terms of discussing the most recent discoveries and advances about the system, this volume focuses on those areas that hold the most promise for furthering scientific understanding and developing new medical treatments for respiratory diseases.

Today's high-tech world offers a wealth of information resources. Programs about the biology of the human body and various diseases are one of the mainstays of educational television programming. The World Wide Web also contains an abundance of information on the human body and the respiratory system. By merely sitting at the computer, anyone can conduct a search and come up with everything from scientific papers to general overviews to simplified articles prepared for grade school students. So why write and publish a volume about the human respiratory system, which is part of the larger *Human Body Systems* series?

First and foremost, this volume represents a compilation of knowledge from a wide range of sources that covers almost every aspect of the respiratory system (as do the other volumes in this series about their specific topics). The information for this work has been carefully researched, sifted through, and presented in an understandable way that is not readily available on television, the Internet, or perhaps in any other modern reference book on the subject. As such, it offers the reader a readily available, easy-to-understand resource to gain an in-depth understanding of the respiratory system, its diseases, and some of the aspects of modern life that greatly affect the system. While a person may be able to go to an Internet search engine, type in "cellular respiration," and come up with pages upon pages of Web sites discussing the topic, the individual would also have to take the time to sift through the various Web pages to find, for example, information written on a level that is understandable by the lay person. It would still take even further comprehensive research to fill in the blanks and gain the overall—yet detailed—view of cellular respiration presented in this volume. Reading this book can also serve as a starting point for those who are interested in learning even more about the system. For this purpose, the volume includes both a bibliography of other resources and a list of organizations and Web sites that the reader can consult for more information.

The human body is fascinating, and the more details discovered about how the body works the more wondrous it becomes. For those readers who are considering a career in the field of science, or those who are just interested in learning how their bodies work, this volume provides a detailed, informative look at the respiratory system and its amazing biological and molecular machinery. Anyone who reads this entire book or just focuses on the chapters discussing how the system works will never take the simple act of breathing for granted again. Breathing is an amazing and intricate process, and the respiratory system is the very foundation of life.

INTERESTING FACTS

▶ At rest, we breathe 15 to 20 times a minute and exchange nearly 17 fluid ounces (about 500 milliliters) of air with each complete breath in and out.

▶ Approximately 5 fluid ounces (about 150 milliliters) of the air we breathe in with each breath fills the passageways of the trachea, bronchi, and bronchioles.

▶ We breathe over 5,000 times a day, taking in enough air throughout a lifetime to fill 10 million balloons.

▶ The average set of human lungs has approximately 600 million alveoli (300 million per lung), creating a respiratory surface about the size of a singles tennis court or a square about 27 to 28 feet long on each side.

▶ At birth, an infant's lung is estimated to have approximately 20 to 30 million alveoli and 1,500 miles of airway passages.

▶ The right lung is slightly larger than the left.

▶ The capillaries in the lungs would extend 1,600 meters or about one mile if placed end to end.

▶ Every minute, 1.3 gallons (5 liters) of blood is pumped through the pulmonary capillaries and around the alveoli.

▶ Overall, blood takes approximately one second to pass through the lung capillaries during which time it becomes nearly 100 percent saturated with oxygen, while losing all of its excess carbon dioxide.

▶ As a result of goblet cells and fine hair-like structures called cilia that help to filter foreign particles out of the air before they can enter the lungs, air breathed in through the nose is cleaner than air entering through the mouth.

▶ We lose nearly 17 fluid ounces (half a liter) of water a day through breathing. This is the water vapor we see when we breathe onto glass.

▶ According to some estimates, a person inhales and ingests approximately 10,000 microorganisms per day.

▶ Researchers estimate that each cigarette shortens your life by approximately 7 to 11 minutes.

Components and Development

From the day we are born and take our first breath, we have set into motion the continuous and essential process of acquiring oxygen (O_2) from the air and eliminating carbon dioxide (CO_2) from the blood. This exchange of gases is called *respiration*. The spontaneous and rhythmic process of breathing is made possible by a complex, finely tuned system of organs, tissues, and passages called the *respiratory system*. Working in conjunction with the cardiovascular system, which pumps 1.3 gallons (5 liters) of blood through the lungs every minute, the respiratory system provides oxygen for the body's cells to produce energy and removes the carbon dioxide waste byproduct created by cellular metabolism.

In addition to **gas exchange**, the respiratory system has other functions. For example, the respiratory tract is lined from the nasal cavities to its smallest branches within the lung with sticky, mucous-secreting cells. These cells help defend the body from environmental pollutants by trapping and eliminating dust, allergy-causing pollens, and other airborne particles. The respiratory system also helps to maintain the body's temperature from 97 to 100°F by releasing warm moist air during exhalation, and it plays a part in balancing the blood's acid-base alkaline composition. Nevertheless, the system's primary function is respiration for gas exchange. There are two distinct modes of respiration (see Chapter 2): organismal (sometimes referred to as external) respiration (involving the lungs) and cellular respiration (involving chemical reactions within the cells). All of the respiratory system's functions begin with a specialized system of structures and organs.

THE COMPONENTS

The respiratory system can be broken down into two portions, each of which performs distinct functions. The conductive portions are composed of structures that act as ducts and pathways connecting the lungs to the outside environment. These include the nasal cavity, pharynx, and other structures. The respiratory portion, which includes the lung and lung structures, facilitates the gas exchange process. In addition, the respiratory system includes ventilating mechanisms, which are the various chest structures and muscles that help to move air in and out of the lungs. The entire respiratory system can also be broken down into two sections: the upper and lower respiratory tracts.

The primary components of the upper respiratory tract are the:

- Nose and nasal cavity (passage)
- Pharynx (throat)
- Larynx (voice box)

The primary components of the lower respiratory tract are the:

- Trachea (windpipe)
- Bronchi
- Alveoli
- Lungs

The Nose and Nasal Passages

Although we sometimes breathe through our mouths (for example, when we run or do strenuous work or when we have a sinus infection), human inspiration (taking in air) and expiration (expelling air) usually takes place through the nose and nasal cavity, which joins the nose and the pharynx. The nasal wall, or **septum**, divides the nasal cavity into two sides. The bottom portion of the nasal cavity is called the hard palate, and three bony ridges or projections, called nasal **conchae**, are on the surface of the cavity sides. The nasal structure also includes the paranasal sinuses (see Figure 1.1). These hollow cavities in the bones of the head connect to the nasal airways via a small passageway in the conchae called a *meatus*. It is unclear what function the paranasal sinuses perform. They may help provide resonance for vocal sounds and lighten the skull. Another theory is that the sinuses may have once aided humans in the ability to smell, as they still do for some lower animals. However, since they no longer perform this function in humans, the paranasal sinuses may be a leftover component that no longer serves an important functional purpose.

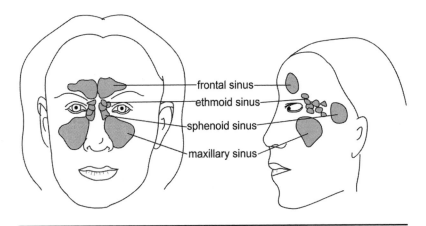

Figure 1.1. Paranasal sinuses.
This diagram illustrates the four basic groups of the paranasal sinuses: frontal, maxillary, ethmoid, and sphenoid sinus.

Air enters through the external openings of the nose, called the *nostrils* or *external nares*. It then passes into the pharynx or throat through interior openings called the *internal nares*. The nasal passages and sinuses between the external and internal nares are lined with mucus-secreting epithelial cells called **goblet cells** and fine hair-like structures called **cilia** (see "Major Cells of the Respiratory System"). Together, these components help to filter foreign particles out of the air before these particles can enter the lungs. This filtering process is achieved when the sticky mucous membrane traps foreign particles, which are then swept by the waving microscopic cilia into the back of the throat, or pharynx, much like seaweed or sea grass buffeted back and forth by the waves. These particles are swallowed and eventually broken down by hydrochloric acid in the stomach and eliminated by the digestive system. The cough reflex can also expel them into the air. As a result of this process, air entering through the nose is cleaner than air entering through the mouth.

The nose and nasal cavities also serve as the body's air conditioner. The nasal passages and mucous membrane warm and humidify air before it enters the lower part of the respiratory system; this function is essential to help prevent harm to other, more fragile linings within the system, such as the lining of the lungs. Several features facilitate this process. Humidification takes place partially because of moisture secreted by the mucous membrane. The nose is also partitioned into two halves by the nasal septum, which is supported by bone and cartilage, thus providing a greater surface area for warming air. **Capillaries** (small blood vessels) that line the nose and cavities also give off heat, and the nasal conchae folds increase surface area and create turbulence that further "conditions" the air.

Major Cells of the Respiratory System

Epithelial cells typically form sheets covering the surface of the body and lining cavities, tubular organs, and blood vessels. They play a major role in the respiratory system. Pseudostratified columnar epithelium cells line the conducting portion of the respiratory tract, from the trachea to the mid-size bronchioles. They are called *pseudostratified columnar* because this sheet of columnar cells (cells that are taller than they are wide) look like they are stratified in layers. However, the "pseudo" prefix means "fake," and these cells are not actually multilayered. Cells making up the pseudostratified columnar epithelium include:

- ciliated cells that have moving cilia to "sweep up" particulate matter

- goblet cells that produce and secrete mucous coverings (primarily in the trachea and bronchi), help humidify the air, and trap foreign particles

- basal cells in the bronchi and bronchioles that may serve as stem, or progenitor, cells to create other cell types, including ciliated and goblet cells

- clara cells that secrete extracellular lining fluid and surfactant proteins

In addition, two essential types of epithelial cells are found in the alveoli:

- Type-I pneumocyte (Alveolar type-I) cells are very thin, flat squamous cells that cover about 95 percent of the alveolar surface and form part of the blood-gas barrier for gas exchange in the alveoli.

- Type-II pneumocyte (Alveolar type-II) cells are situated at the junctions between alveoli and synthesize and secrete phospholipid-rich surfactant; they also proliferate in response to lung injury acting as a progenitor, or precursor, for the type-I cells.

Pharynx

The pharynx, commonly referred to as the throat, is the funnel-shaped opening leading from the nose and mouth to both the lower respiratory tract and the digestive system. While food passes through the pharynx into the esophagus and stomach for digestion, air passes through the nose and pharynx and on into the larynx and trachea, which leads directly into the lungs (see Figure 1.2). The pharynx, which is about five inches long, is typically divided into three segments called the *nasopharynx* (upper), *oropharynx* (middle), and *laryngopharynx* (lower). (The nasopharynx serves exclusively as part of the respiratory tract. The oropharynx and laryngopharynx also help guide food into the alimentary tract. On swallowing, a muscular flap called the soft palate closes the nasopharynx off from the oropharynx. The

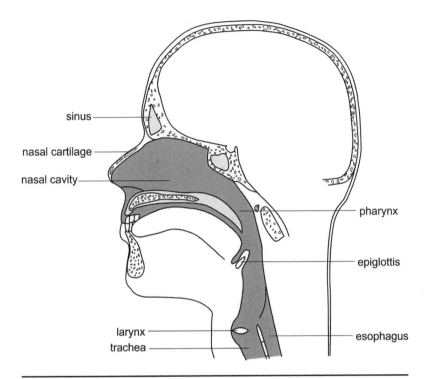

Figure 1.2. Upper respiratory tract.
This diagram shows the upper respiratory tract with the upper portion of the trachea and esophagus included.

laryngopharynx connects the oropharynx with the esophagus.) Lined with mucous secreting epithelial cells to help remove foreign particles, the pharynx also helps to warm and humidify air before it reaches the lungs.

Larynx

Composed of bone, cartilage, and muscle, the larynx is a valve-like structure that separates the trachea from the upper respiratory tract and connects the pharynx and trachea. It includes the large thyroid cartilage that can protrude from the front of the neck, commonly called the "Adam's apple." Although often referred to as the "voice box" because it contains the vestibular (vocal) folds and chords needed for human speech, the larynx serves important regulating functions during respiration. Both the vestibular folds and the epiglottis, a flap-like tissue composed of elastic cartilage that sits above the larynx, act similar to trap doors that open to allow air to enter and close to prevent aspiration (food from entering the lower respiratory tract). The larynx, which is also lined with mucosal epithelium, also helps the respiratory system rid itself of impurities through the coughing mechanism activated by nerves that are extremely sensitive to touch. Laryngitis develops when mucosal epithelium on the vocal chords become inflamed.

Trachea

The trachea, commonly referred to as the windpipe, is a tube-like structure stabilized by fifteen to twenty C-shaped pieces of cartilage. It is typically 4 to 5 inches (10 to 12 centimeters) long and around one inch (2.5 centimeters) in diameter. In addition to serving as the primary passageway of air into the lungs, the trachea contains mucus-producing epithelium to trap foreign particles. Cilia are also present to propel these particles upward toward the larynx for swallowing or expiration. As the lower end of the trachea enters the lungs, it branches off behind the sternum (breastbone) into the left and right primary **bronchi**, which enter the left and right lung. Because the right bronchus is shorter, wider, and more vertical than the left bronchus, food usually enters the lower respiratory tract via the right bronchus when it bypasses the esophagus and "goes down the wrong pipe."

Bronchi and Bronchioles

The primary bronchi are similar to the trachea in that they also have an epithelium lining and are supported by C-shaped cartilage. The final portion of the respiratory system's conductive segment, the primary right and left bronchi branch off further into increasingly smaller bronchi down to approximately 0.04 inches (1 millimeter) in diameter. These differ in construction from the primary bronchi in that their support comes from smaller cartilage plates embedded in the walls.

The complex system of bronchi that branches throughout the lungs is called the bronchial tree, which extends further and further into finer "branches" that have less cartilage for support and more smooth muscle. These ultimately become **bronchioles**, which are approximately 0.02 inches (0.5 millimeters) in diameter. The division leads to terminal bronchioles and then respiratory bronchioles. These tiny tubes, which further divide to form alveolar ducts that will end in air sacs called **alveoli,** are considered the first structures that belong to the respiratory portion rather than the conductive portion of the respiratory system (see Figure 1.3).

Alveoli

The respiratory bronchioles end in small grape-like clusters of alveoli (individually called *alveolus*), where the gas exchange between oxygen and carbon dioxide takes place. As bronchioles continue to divide, the number of alveoli increases. The average set of human lungs has approximately 600 million alveoli (300 million per lung), creating a respiratory surface in the vicinity of 750 square feet (about 70 square meters, which is about the size of a singles tennis court or a square about 27 to 28 feet long on each side).

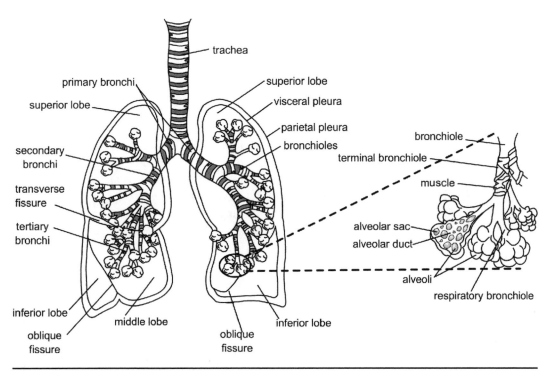

Figure 1.3. Lower respiratory tract.
This diagram shows the major passages and structures of the lower respiratory tract.

Each individual alveolus is approximately 0.004 to 0.007 inches (0.1 to 0.2 millimeters) in diameter. Like tiny balloons, the alveoli inflate and deflate. Alveoli have thin, one-celled walls made of squamous epithelial cells (epithelial cells with a scaly outer layer) that are covered by an extensive network of fine capillaries. Each single alveolus is surrounded by about 2,000 segments of capillaries, which have single-layered endothelial cell walls. Gas exchange occurs via diffusion (net movement of particles from a region of higher concentration to a region of lower concentration) between the thin walls of the alveoli and the capillaries. The blood-gas barrier, or respiratory membrane, has a thickness of approximately one-half of one micrometer (a micrometer is 1/1000 the thickness of a dime). The process involves oxygen passing from alveoli into capillaries for distribution throughout the body, and carbon dioxide diffusing from the capillaries into alveoli where the gas is eliminated through expiration.

A fluid called a **surfactant** is produced by type-II pneumocytes (specialized cells that line the alveoli) and secreted in the alveoli to coat the walls and reduce surface tension or stiffness. Reduction of surface tension results in less pressure being needed to inflate the alveoli, which is especially important at birth. Surfactant lining the alveoli also provides the moist surface necessary for gas exchange, because gas must dissolve in liquid before moving through cells. As a substance, surfactant has a half-life of fourteen to twenty-eight hours, meaning that it degrades very quickly and must be continually produced by the pneumocytes.

Lungs

The bronchial tree and alveoli course throughout the conical-shaped left and right lungs. In terms of volume, the lungs are one of the largest organs of the body and the two lungs together weigh a total of approximately 1.7 and 2.2 pounds (between 800 and 1,000 grams). They take up the majority of chest space (thorax), which comprises the space from the base of the neck to the diaphragm, upon which the lungs sit. The slightly larger right lung has three lobes (the superior, middle, and inferior lobes), and the left lung has two (the superior and inferior lobes). Deep fissures, or crevices, on the lung's surface define the separate lobes.

Each lung is enveloped by a transparent membrane called the **pleura,** which has an outer membrane (parietal pleura) attached to and lining the thoracic, or chest, wall and an inner membrane (visceral pleura) that tightly covers the lungs. Between the outer and inner pleural membranes, which are actually one continuous membrane that doubles back to cover both the chest and lungs, is a space called the pleural cavity. Inside this cavity is the pleural fluid, which helps to reduce friction between the membranes during breathing when the lungs expand and contract and also helps hold both pleural layers in place, much like two microscopic slides that are wet and stuck together. The lungs are also encased by the rib cage, which provides protection from outside trauma. Between the right and left lungs is an area called the *mediastinum*, which contains the heart, trachea, esophagus, thymus, and lymph nodes. The heart separates the right and left lung, and the left lung's smaller size includes a "cardiac notch" to provide space for the heart to extend into.

Diaphragm and Intercostal Muscles

The diaphragm and the **intercostal muscles** are the ventilating mechanisms, or muscles, that allow the lungs to bring in and expel large volumes of air. At rest, we breathe fifteen to twenty times a minute and exchange nearly 17 fluid ounces (about 500 milliliters) of air with each complete breath in and out. The dome-shaped diaphragm is attached to the lower six ribs via a central tendon. The intercostal muscles line the rib cage, with the external intercostals running forward and downwards and the internal intercostals running upwards and back. Together, they form sheets that stretch between successive ribs. The diaphragm and intercostal muscles

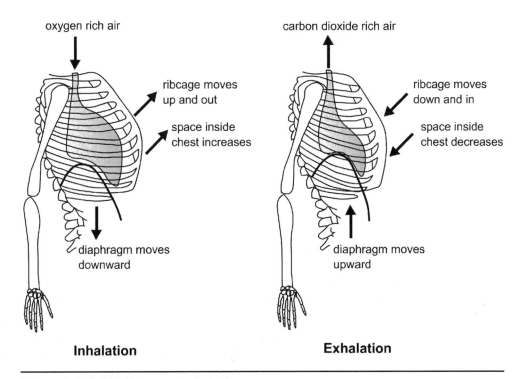

oxygen rich air

carbon dioxide rich air

ribcage moves
up and out

space inside
chest increases

diaphragm moves
downward

ribcage moves
down and in

space inside
chest decreases

diaphragm moves
upward

Inhalation   **Exhalation**

Figure 1.4. Inhalation and exhalation.
This diagram shows the movement of the diaphragm and ribcage during inhalation and exhalation.

help the chest area and the lungs to expand and contract. During inspiration, the external intercostal muscles contract and lift the ribs up and out, and the diaphragm contracts. This process increases the size of the chest cavity and reduces air pressure inside the lungs compared to the air outside, creating a vacuum that inflates and draws air (via the trachea) into the lungs. During expiration the process is reversed, with the intercostal muscles moving the ribs downward and the diaphragm moving up creating a smaller chest cavity that increases lung pressure and forces air out (see Figure 1.4). The natural elasticity of the lungs helps return them to their normal volume.

DEVELOPMENT

The respiratory system begins to develop in the embryonic stage of life as cells start to divide. It continues postnatally (after birth) for at least two years and possibly for as long as ten years. This multi-event process involves more than forty different cell types that differentiate (specialize in structure and function) and proliferate.

Prior to birth, the human fetus acquires oxygen and eliminates carbon dioxide via the mother's bloodstream. However, the blood of mother and fetus never mix, so the mother's red blood cells are not exchanged with the fetus's. Rather, the umbilical cord contains fetal **arteries** and **veins** that carry diffused molecules between the placenta and the fetal circulatory system. Oxygen molecules released by **oxyhemoglobin** (**hemoglobin** with oxygen bound to it) in the mother's blood migrate through the placenta. The oxygen molecules then are picked up by the hemoglobin in the fetus's red blood cells and carried throughout the fetus's body. This process is enhanced by the fetal blood's greater oxygen-carrying capacity compared to maternal blood.

Although the respiratory system does not perform its physiological function of gas exchange until after birth, the respiratory tract, diaphragm, and lungs begin forming early in embryonic development at approximately four weeks of gestation. However, once it reaches the fetal stage at approximately seven to eight weeks, the fetus does move its lung muscles about 40 percent of the time. This fetal breathing movement takes amniotic fluid (which bathes the fetus and is primarily composed of fetal urine containing electrolytes and protein) into the lung and is essential to conditioning the lung for postnatal breathing by providing a stretch to lung tissue. It stretches lung tissue by creating a pressure gradient between the lungs and the amniotic fluid compartment, which also creates an outflow resistance in the trachea. This resistance keeps the alveoli and airways distended.

Differentiated tissue for all the body systems and components develop from three primordial germ cell layers formed in the embryo during its early development. These are the ectoderm (outermost layer), mesoderm (middle layer), and endoderm (innermost layer) of the forming embryo. The respiratory system develops primarily from the mesoderm and endoderm. The endoderm germ layer differentiates into the larynx, trachea, and lung, and ultimately the lining of the respiratory tract. The mesoderm gives rise to the vascular system necessary for transportation of oxygen, as well as to other connective tissues, lymphatics, bone, and cartilage throughout the body.

The respiratory system's development begins at approximately the same time that the human embryo changes from a flat disc into a three-dimensional embryo (see "The Molecular Basis of Development"). At this point, the body begins a process called *lateral body folding,* during which time the endoderm forms a gut tube. When the body folding is completed, the gut tube is divided into three segments: the foregut, midgut, and hindgut. The ensuing respiratory system development begins as an endodermal outgrowth from the foregut. This outgrowth eventually differentiates into the pharynx, trachea, bronchi, and lungs.

The Molecular Basis of Development

Although much is understood about the respiratory system's development, scientists are fine-tuning their understanding by delving into the molecular and genetic factors that play such a vital role in the process of development. For instance, timing is essential during organ and system development in the embryo and fetus. Researchers are learning about a variety of factors that control the timing and pattern of cellular proliferation, migration, and differentiation. For example, transcription and growth factors, proteins, and various molecular signals are expressed in specific temporal and spatial patterns to ensure normal lung development, including cell proliferation, cell-to-cell interactions, and even branching of the bronchi.

While identifying factors in the respiratory system's development, scientists are sometimes surprised by what they find. In 2000, researchers at the Massachusetts Institute of Technology (MIT) discovered that a chemical receptor found in the brain—the N-methyl-D-aspartate (NMDA) receptor—may be key to fetal development of the respiratory system. This receptor has been closely associated with learning and memory. When scientists "knocked out" this receptor during prenatal development to see how it affected learning and memory in mice, they discovered that it also led to fatal respiratory distress.

The development of the respiratory system can be broken down in five stages: the embryonic, pseudoglandular, canalicular, saccular, and alveolar phases. Estimates of time frames for these stages may vary by a week or two.

Embryonic Phase (3 to 6 Weeks)

Typically, the unborn child or offspring is called an embryo during the first six weeks of development. At approximately four weeks, the **laryngotracheal groove** forms in what is the floor of the primitive pharynx. As the groove deepens, it forms the laryngotracheal diverticulum, which eventually separates from the foregut in the form of tracheoesophageal (relating to the trachea and esophagus) folds. These folds fuse to form the tracheoesophageal septum and ultimately the esophagus and laryngotracheal (relating to the larynx and trachea) tube. A single lung bud also forms from the foregut at the tail end of the laryngotracheal groove. At the end of four weeks, or approximately day 26 to 28 of gestation, the lung bud begins to separate into right and left bronchial buds. At week six, segmental bronchi form to establish the bronchopulmonary (relating to the bronchial tubes and the lungs) segments, marking completion of the proximal (near the point of origin or attachment) airway formation. Development of the **pulmonary** vasculature also occurs at this time as it penetrates into the primitive mesenchyme (a loose network of undifferentiated mesoderm cells). At this point, the primitive lungs begin to expand.

Pseudoglandular Phase (7 to 16 Weeks)

From this point until birth, the child is considered to be in the fetal stage. As this phase begins, the lungs look like a gland. Conducting airways begin to form and rapidly divide into successive generations of branching bronchial buds, reaching around seventeen to twenty divisions by the end of sixteen weeks, and twenty-four generations by early childhood. With each division, air passages become narrower while the number of airways increases geometrically. The last eight generations of branching form the terminal bronchioles. Fetal breathing movements begin at approximately eleven weeks. At eleven to thirteen weeks cilia appear in the proximal airways. By week sixteen, most of the bronchial airways are formed. As this stage ends, capillaries form new channels (canalize) in the forming lung tissue (parenchyma). The respiratory airways and vasculature soon correspond physically—but not functionally—to those of an adult. In essence, all major components of the pulmonary system have formed except for those necessary for gas exchange, such as the alveoli.

Canalicular Phase (17 to 26 Weeks)

Over this ten-week period, lung morphology (form and structure) changes dramatically as the basic structure of the lung's gas-exchanging portion is formed and vascularized. During this stage, the terminal bronchioles branch to form several orders of primitive respiratory bronchioles. The respiratory bronchioles also contain terminal sacs that will eventually give rise to alveoli. Vascularization of the mesoderm associated with the respiratory bronchioles also occurs. Surfactant production begins at approximately twenty-four weeks. It is during this stage that the developing gas exchange regions of the lungs can be easily distinguished from the developing conducting airways of the lungs. By the end of the canalicular stage, the fetus possibly could breathe and survive outside the womb.

Saccular (or Terminal Sac) Phase (27 to 36 Weeks)

At this stage, alveoli begin to develop as still primitive pouches (saccules) in the walls of the respiratory bronchioles. Secondary dividing walls or membranes (septa) also grow into the airspaces, partitioning the more primitive pouches into alveolar ducts (enlarged terminal sections of bronchioles) and sacs. A fully mature alveolar is akin to a room with one wall missing, which creates a passageway into the alveolar ducts and sacs. The vascular "tree" also grows in length and diameter during the saccular stage.

Alveolar Phase (36 Weeks to Term/Birth)

The mature alveoli begin to appear at thirty-six weeks and continue to grow and mature after birth. The membranes that separate the alveoli from the capillaries become extremely thin, forming a rudimentary but still functional blood-gas barrier. Prior to birth, the alveoli are partially filled with fluid. Nevertheless, the full-term infant's lungs have smaller, immature alveoli marked by a thick septa, as compared to the alveoli of adults.

Postnatal Period

A newborn infant's first breath must be forceful to initially inflate the lungs. As the baby passes through the birth canal, much of the fluid in the alveoli is squeezed out and replaced with air. The expelled fluid is absorbed in lung tissue during the first few hours after birth and eventually cleared by lymphatic structures and pulmonary blood flow.

At birth, the full-term lung is estimated to have approximately 20 to 30 million alveoli and 1,500 miles of airway passages. Thus, approximately 80 to 90 percent of alveoli are formed postnatally, most before age 3. The transition from bilayered, primitive septa between the alveoli to the thin mature septa containing only a single capillary layer also begins to take place shortly after birth. Complete respiratory development, including enlargement of the lung due largely to the multiplication and increased size of alveoli, can take up to eight to twelve years.

Developmental Abnormalities

Various malformations of the respiratory system can occur during their development, including tracheoesophageal fistulas (abnormal passages from the trachea to the esophagus that per-

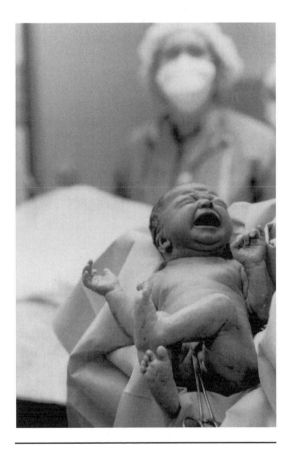

A newly delivered baby takes its first breaths. Although a fetus does make breathing motions and moves amniotic fluid in and out of the lungs, true breathing begins just after birth as the baby takes its first breath of air. © Hulton/Getty Archive.

mit passage of fluids or secretions), pulmonary hypoplasia (immature pulmonary development), respiratory distress syndrome, diaphragmatic hernias, and esophageal atresia (absence of the normal continuous passage through the esophagus). Respiratory distress syndrome is the leading cause of death in newborns, occurring in approximately one percent of pregnancies and affecting 30,000 newborns each year in the United States. The disease is due to insufficient or no surfactant production, which makes it difficult or impossible for the alveoli to expand and the newborn to breathe.

The potential for developmental problems does not end at birth. After birth, the developing lungs may also be especially susceptive to toxic environmental agents, including environmental, or "secondhand," tobacco smoke and oxidant gases, such as ozone and nitrogen dioxide. These problems and other respiratory diseases will be discussed in more detail in Chapters 4 and 5.

Respiration

We are taught at an early age that eating the proper foods is essential for good health. But without the respiratory system the food we eat could not sustain us. The oxygen (O_2) supplied by the respiratory system plays a fundamental role in enabling the body's cells to turn food into life-producing energy.

Beginning with nearly 2 gallons (approximately 6 to 7 liters) of fresh air we breathe in every minute, the respiratory system inhales close to 3,000 gallons of air each day to acquire the oxygen necessary to fuel the metabolic processes that create energy from the carbohydrates found in food. This energy enables the body's cells to multiply and function. The process also results in the generation of the waste product carbon dioxide (CO_2), which the respiratory system helps to eliminate. If we did not breathe in fresh oxygen, carbon dioxide would rapidly accumulate to toxic levels within the blood and result in death. As discussed in Chapter 1, the process of bringing in oxygen from the atmosphere and eliminating carbon dioxide is called *gas exchange*.

Although the word *respiration* comes from Latin meaning "to breathe again," respiration is much more than merely "breathing" air in and out. In the human body, respiration encompasses many processes, from the readily perceptible act of breathing to the hidden, complex metabolic machinery that continuously works within the body's trillions of individual cells. In terms of human biology, respiration operates on two basic levels. The first is called *organismal respiration* and refers to the entire human body, or "organism," taking in oxygen from the environment and returning carbon dioxide to it. The second is called *cellular respiration* and encompasses the metabolic activities that occur when the body's cells use oxygen and food to generate energy and produce carbon dioxide.

ORGANISMAL RESPIRATION

Organismal respiration involves four stages:

- Pulmonary ventilation (movement of air in and out of the lungs)
- External respiration (gas exchange between the lungs and the blood)
- Internal respiration (gas exchange between the blood and the body's cells [tissues])
- Transportation (movement of oxygen and carbon dioxide through the body via the blood)

Pulmonary Ventilation and the Mechanics of Breathing

The respiratory system's function begins with the exchange of large volumes of air between the environment and the lungs via inspiration and expiration. This process is referred to as *pulmonary ventilation.* Although pulmonary ventilation primarily serves to bring oxygen into the body from the atmosphere and as the final stage of expelling carbon dioxide waste from the body, it is during this initial process that the respiratory system performs many of its functions secondary to gas exchange (see "How Do Other Animals Breathe?"). Air is warmed and moistened as it enters the nose, which helps to maintain our body temperature. The respiratory system also filters out environmental pollutants, such as dust. For example, mucus secreted by goblet cells lining the airways and lungs traps particles and then cilia sweep the mucus upwards from the throat for swallowing or expulsion via coughing. The respiratory system also helps to balance our body's acid, or **pH**, levels through its role in regulating the elimination of carbon dioxide. Control of pH is essential for the proper functioning of enzymes, proteins, and other biological processes.

As with all the major systems of the human body, the respiratory system works in conjunction with other major systems. In the case of pulmonary ventilation, the nervous system exerts initial control over the breathing process, including the rhythm, rate, and depth of breathing. The message centers in the brain that control rhythmic respiration of breathing in and out are located in the brain stem and are called the *pons* and the *medulla oblongata* (see Figure 2.1). These autonomic (or automatic) brain centers are more primitive than parts of the brain located in the cortex, which give us control over our movements and thoughts.

Whether we think about it or not, the pons and the upper portion of the medulla automatically regulate our breathing, which is why we can still breathe while we sleep. However, we can exert voluntary control over our

How Do Other Animals Breathe?

Although respiration is necessary for life, the respiratory process varies greatly depending on the life form. A single-celled organism, such as an amoeba or paramecium, achieves gas exchange via diffusion across the cell membrane. Because of oxygen's slow diffusion rate as compared to carbon dioxide, organisms that achieve respiration in this manner are extremely limited in size. Smaller animals that do not have specialized gas exchange organs also rely on diffusion. In the earthworm, oxygen dissolves on the worm's moist skin, diffuses into the capillaries just beneath the skin, and then is carried throughout the body by hemoglobin. In turn, carbon dioxide in the blood diffuses out from the capillaries, through the worm's skin, and into the air. To acquire enough oxygen, these animals must have a high surface-to-volume ratio and, as a result, are often flat and tubular in shape. Sponges and jellyfish also perform gas exchange via their entire surfaces, taking in oxygen directly from the water around them.

Because larger animals are more complex with many more cells that require oxygen, they cannot acquire enough oxygen by diffusion across their outer surfaces. As a result, they develop specialized respiratory surfaces that provide increased surface area for gas exchange. Crustaceans, fish, and some amphibians have gills, either internal (crabs and fish) or external (some amphibians). These gills are typically branched in a way that resembles feathers to increase surface area. Gills contain many capillaries, covered by a thin layer of epithelial cells. As water flows over them in one direction, blood flows through the capillaries in the opposite direction to further maximize oxygen transfer into the animal's body. (Oxygen transfer in aquatic animals must be extremely efficient, because water contains about one-twentieth the amount of oxygen as the same volume of air.) Other animals, including many insects, have tiny openings called *spiracles* on their body's surface. These openings lead to numerous air tubes, similar to the human trachea, that transfer air directly from the atmosphere to cells, with the rate of diffusion depending on body movements or contractions.

breathing when needed. For example, we can hold our breaths for a certain length of time under water or consciously make ourselves breathe faster or slower. The part of the brain that allows conscious control of breathing is located in the cerebral cortex.

During automatic respiration, specific **neurons** in the medulla and pons send signals to motor neurons in the spinal cord, approximately ten to twelve times each minute. These nerve cells, in turn, signal the diaphragm and intercostal muscles surrounding the thoracic cage to contract, thus expanding the rib cage and lungs within it. When the lungs expand, pressure within the lungs becomes lower than the pressure in the atmosphere, causing air to rush in through the conductive portion of the respiratory system

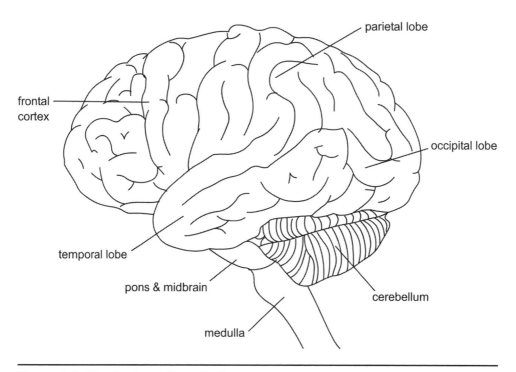

Figure 2.1. The brain and respiration.
This figure shows the location of the pons and the medulla, which are the parts of the brain that automatically control our rhythmic breathing.

(nose, pharynx, larynx, trachea, and bronchi) until full expansion is reached. The air we inhale contains approximately 21 percent fresh oxygen and little or no carbon dioxide.

When the lungs are fully expanded, the **vagus nerve** tells the brain to "turn off" its signals for inspiration. As a result, the muscles surrounding the thoracic cage relax, moving the ribs back to their resting state. At this point, expiration is passive in that the brain and the muscles surrounding the lungs do not directly regulate it. The resulting decrease in chest cavity size contracts the lungs, causing us to exhale because the space available for air in the lungs is reduced. In other words, the air pressure in the lungs becomes higher than the pressure in the atmosphere. The resulting exhaled "stale" air is made up of about 16 percent oxygen and 6 percent carbon dioxide.

Although pulmonary ventilation is primarily automatic because we do not consciously control our breathing most of the time, the brain is actually hard at work interpreting **neurochemical** information to control breathing, including monitoring respiratory volume and blood gas levels. Surprisingly, the primary stimulus for the brain to control breathing is not the amount of oxygen in the blood, but rather the amount of carbon dioxide. Chemical receptors, or **chemoreceptors,** in the medulla, in collaboration with chemoreceptors in the **carotid arteries** and the **aorta**, respond to carbon dioxide levels in the blood. High carbon dioxide concentrations result in deeper and faster breathing designed to bring in higher levels of oxygen and reduce harmful carbon dioxide levels. In turn, our respiration rates slow down when carbon dioxide levels are lower. However, oxygen levels can also affect respiration. When the **aortic bodies** and **carotid bodies** detect low levels of oxygen, they send signals to the brain stem to make breathing more rapid and deeper.

Other factors affect the brain's regulatory function of respiration.

- An increase in blood pressure slows respiration.

- A sudden decrease in blood pressure increases respiration.

- A decrease in blood acidity (higher pH levels) increases respiration. (This state usually results from oxygen debt, or a lack of oxygen reaching the muscles, which produces lactic acid and lowers the pH level.)

Although some diseases can affect the body's oxygen and carbon dioxide levels, these levels are most commonly affected by physical activity. For example, hard work or exercise, especially when **aerobic** in nature, causes the body's cells to metabolize faster to create more energy. More carbon dioxide is produced in the process and eliminated into the blood, thus lowering the blood's pH level. As a result, during physical exertion the body can increase oxygen consumption up to twenty-five to thirty times more than when the body is at rest.

External Respiration

External respiration is the exchange of oxygen and carbon dioxide between the lungs and circulating blood. Just as the respiratory system works in conjunction with the nervous system during pulmonary ventilation, it also works with the heart and circulatory system to pump blood to the lungs during external respiration. This process is called *pulmonary circulation.* The movement of blood away from the lungs and heart to other parts of the body is called *systemic circulation.*

The amount of air we breathe in and out of the lungs during pulmonary ventilation is called *tidal volume.* Although we breathe in about one pint (around 500 milliliters) of air with each breath, approximately .32 pints (150 milliliters) of this air fills the passageways of the trachea, bronchi, and bronchioles. When filled with tidal volume air, these conductive portions of the respiratory system are referred to as *anatomical dead space*, meaning that the air remaining in this space is not involved in the external respiration process. The air that passes through the last conductive portions of the respiratory system enters the millions of alveoli in the lungs. It is here that external respiration takes place when oxygen and carbon dioxide are exchanged between air in the alveoli and the minute blood vessels called *pulmonary capillaries* that surround each individual alveolar sac like a net.

Every minute, 1.3 gallons (5 liters) of blood is pumped through the pulmonary capillaries and around the approximately 600 million alveoli in the lungs. The air-filled alveoli contain more oxygen compared to the blood in the capillaries. Conversely, blood in the capillaries contains more carbon dioxide than air in the alveoli. As a result, the exchange of oxygen and carbon dioxide between the capillaries and the alveoli occurs via diffusion across the microthin membrane walls separating the two.

The diffusion of all gases primarily depends on their solubility in water and their **partial pressure**, which expresses the concentration of a gas. The concentration, or diffusion gradient, of the gas is expressed in **millimeters of mercury** (mmHg). (For example, the partial pressure of oxygen would be expressed as pO_2 # mmHg.) Because the random movement of oxygen and carbon dioxide molecules results in their net movement from a region of higher concentration to a region of lower concentration, the concentration or partial pressure of oxygen in the alveoli must be kept at a higher level than in the blood. Likewise, the concentration of carbon dioxide in the alveoli must be kept at a lower level than in the blood. These different levels are maintained because the continuous inspiration of fresh air supplies an abundance of oxygen to the lungs and alveoli, while expiration eliminates carbon dioxide from the body into the air.

The blood carrying the carbon dioxide travels from all parts of the body into the heart's **right atrium** and then into the **right ventricle**, where it is pumped into the **pulmonary artery.** This artery, which is the only artery in the body that carries deoxygenated blood, branches into the right and left lung, ultimately feeding blood into the pulmonary capillaries. Blood entering the capillaries surrounding the alveoli has a pCO_2 of 45 mmHg and a pO_2 of 40 mmHg. Conversely, the environmental air that has entered the alveoli during inspiration has a pCO_2 of 40 mmHg and a pO_2 of 100 mmHg. As a result, oxygen diffuses across microthin membranes into the blood from the alveoli and carbon dioxide diffuses into the alveoli from the blood (see Figure 2.2).

When a person exhales, the air in the alveoli is breathed out into the atmosphere along with the abundance of carbon dioxide that it now contains. Conversely, the oxygen-rich blood is pumped throughout the body via the **systemic capillaries** to the cells that make up various tissues. Overall, blood takes approximately one second to pass through the lung capillaries, during which time it becomes nearly 100 percent saturated with oxygen while losing all of its excess carbon dioxide.

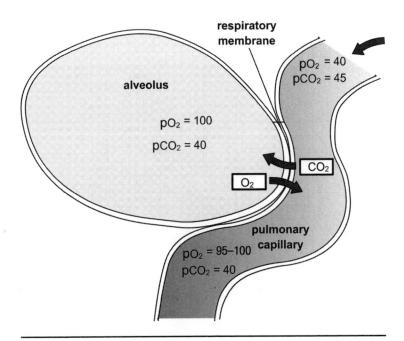

Figure 2.2. Gas diffusion from alveolus.
This diagram shows that carbon dioxide diffuses from the blood into the alveolus, where the concentration of carbon dioxide is much lower, and that oxygen in the alveolus diffuses into the pulmonary capillary, where the concentration of oxygen is lower. Once this occurs, the oxygen-enriched blood flows to the body's tissues and cells.

Internal Respiration

Although internal respiration is sometimes used in the same sense as cellular respiration to refer to the metabolic process within the cells, it is most often used to designate the gas exchange process between blood in the capillaries and the body's cells. Once the external respiration process is completed, the oxygenated blood travels from the alveoli to heart's **left atrium**. The blood moves to the heart's **left ventricle**, then is pumped throughout the body via a network of arteries that feed the capillaries surrounding the body's various tissues. As mentioned earlier, this is known as systemic circulation.

When the blood returning from the lungs reaches the tissues, it has a pO_2 of 95–100 mmHg and a pCO_2 of 40 mmHg. Conversely, the cells that make up our tissues have a pO_2 of 30–40 mmHg and a pCO_2 of approximately 45 mmHg, depending on the metabolic activity within the cell. Again, since the diffusion of gases occurs from an area of higher concentration to lesser concentration, oxygen from the blood diffuses across the interstitial fluid (liquid found between the cells of the body) and into the cells,

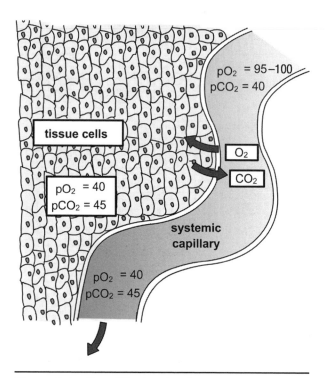

Figure 2.3. Gas diffusion in tissues.
This diagram shows that oxygen-enriched blood (carried from the alveoli and the pulmonary capillary via the systemic capillaries) diffuses into tissue cells, which have a lower concentration of oxygen. At the same time, the higher concentration of carbon dioxide in the tissues diffuses into the systemic capillary for eventual transport to the alveoli.

or tissues. Conversely, the carbon dioxide from the cells and tissues diffuses into the blood (see Figure 2.3).

Transporting the Gases

The gas exchange process necessary for cells to function properly could not occur without the blood transporting oxygen and carbon dioxide throughout the body. This transportation depends on the gases' distinct properties and on a blood component called hemoglobin, an oxygen carrying **protein** found in red blood cells called **erythrocytes**.

O_2 TRANSPORT

Compared to carbon dioxide, oxygen is not very soluble. As a result, only about 0.01 fluid ounces (0.3 milliliters) of oxygen will dissolve in every 3.4 fluid ounces (100 milliliters) of blood **plasma**, which is not enough to carry sufficient oxygen to the body's tissues and cells. The majority of oxygen in the human body is carried via hemoglobin, which is the respiratory pigment in humans and also gives blood its red color. Because of the affinity of oxygen to hemoglobin, the oxygen carrying capacity of blood is boosted nearly seventy-fold to about 0.7 fluid ounces (20.8 milliliters) per 3.4 fluid ounces (100 milliliters) of blood.

Hemoglobin's unique molecular characteristics make it an excellent transport molecule for oxygen. Each hemoglobin molecule includes four **hemes**, which are iron-containing **porphyrin** compounds, combined with the protein globin. Porphyrins are a group of organic pigments characterized by a ringed group of four linked nitrogen-containing molecular rings (called a *tetrapyrrole nucleus*). In a heme, each porphyrin ring has an atom of iron (Fe) at its center. Each iron atom can unite with one molecule of oxygen. As a result, each hemoglobin molecule can carry four oxygen molecules. Furthermore, when one oxygen molecule binds to one of the four heme groups, the other heme groups change shape ever so slightly so that their affinity increases for the binding of each subsequent oxygen molecule. In other words, after the first oxygen molecule is attached, the next three oxygen molecules attach even more rapidly to form oxyhemoglobin (the bright red hemoglobin that is a combination of hemoglobin and oxygen from the lungs), thus providing rapid transfer of oxygen throughout the blood. Conversely, when it comes time for hemoglobin to "unload" its oxygen content into cells and tissues, once one heme group releases its oxygen, the other three rapidly follow.

Oxygen's affinity for hemoglobin is also affected by the partial pressure of carbon dioxide and the blood's pH level. This is known as the **Bohr effect**, named after its discoverer Christian Bohr (1855–1911). A high concentration, or partial pressure, of carbon dioxide makes the blood more acidic, which causes hemoglobin to have less affinity for oxygen. As a result, in tissues where the concentration of carbon dioxide in the blood is high because of its release as a waste product from cells, hemoglobin easily releases oxygen. In the lungs, where blood carbon dioxide levels are low because of its diffusion into the alveoli, hemoglobin readily accepts oxygen.

The Bohr effect or shift, which relates to a mathematically plotted curve called the **oxygen dissociation curve,** serves an extremely useful purpose. During exercise, cells are working harder—more actively respiring—to produce more energy. As a result, they release much higher levels of carbon dioxide into the blood than when the body is at rest. The higher carbon dioxide levels, in turn, reduce the blood's pH level, thus acidifying the blood and signaling hemoglobin to release more rapidly the oxygen needed to replenish cells and tissues. In other words, the Bohr effect informs the body that its metabolism has increased due to exercise and that it must compensate for the increased need to absorb oxygen and release carbon dioxide.

CO₂ TRANSPORT

Carbon dioxide enters the blood as a waste product of cell metabolism and cellular respiration. Unlike oxygen, carbon dioxide readily dissolves in blood. Carbon dioxide is transported by the blood to the alveoli in three ways:

1. as soluble CO_2 in blood (5 percent–10 percent)
2. bound by hemoglobin (20 percent–30 percent)
3. as a bicarbonate (60 percent–70 percent)

Although carbon dioxide is more soluble than oxygen and dissolves directly into the blood after it diffuses out of cells, the amount that dissolves is not enough to perform the essential function of ridding the body of carbon dioxide. In the second mode of transport, approximately a quarter of the carbon dioxide eliminated from cells reacts with hemoglobin. In essence, carbon dioxide is able to hitch a ride with hemoglobin because, at this point, hemoglobin is not carrying much oxygen and has an increased affinity for carbon dioxide. This is known as the **Haldane effect** and occurs as blood passes through the lungs. Blood proteins that bind to carbon dioxide are called **carbamino compounds.** When carbon dioxide binds to the hemoglobin's protein, the combination is called *carbamino-hemoglobin.*

The first two methods of transporting carbon dioxide are relatively slow and inefficient compared to the third method of transporting the gas. Because carbon dioxide is highly soluble, it reacts readily with water (H_2O) molecules to form carbonic acid (H_2CO_3) in red blood cells. This reaction would also be too slow for efficient carbon dioxide transport if not for an **enzyme** called carbonic anhydrase (CA), which is highly concentrated in red blood cells and acts as a catalyst to help produce carbonic acid. The carbonic acid then ionizes (or disassociates) to form a positively charged hydrogen **ion** (H+) and a negatively charged bicarbonate ion (HCO_3_). The chemical process can be viewed as follows:

$$CO_2 + H_2O \longleftrightarrow H_2CO_3 \longleftrightarrow H+ + HCO_{3\text{-}}$$

Because the concentration of the negatively charged bicarbonate ions in the red blood cells is at a higher level than outside of these cells, these ions readily diffuse into the surrounding blood plasma for transport to the alveoli. To compensate for the negatively charged bicarbonate ions moving out of a red blood cell, a negatively charge chloride (Cl) ion enters the cell from the plasma to maintain the electrical balance in both the erythrocyte and the plasma. This exchange is called a **chloride shift**.

CELLULAR RESPIRATION

Cellular respiration is the process by which cells use the oxygen delivered by the respiratory and circulatory systems to manufacture and release the chemical energy stored in food, primarily in the form of carbohydrates. As such, it is called an *exergonic reaction*, meaning that it produces energy. Cellular respiration produces energy via a catabolic process, that is, by making smaller things out of larger things. In cellular respiration, it refers to the breaking down of polymers (large molecules formed by the chemical linking of many smaller molecules) into smaller and more manageable molecules.

The catabolic process within cells involves breaking down glucose, a simple sugar in carbohydrates that stores energy, into smaller molecules called *pyruvic acid*. These smaller molecules are ultimately used to produce **adenosine triphosphate (ATP)**. ATP is the primary "energy currency" of the cell, the human body, and nearly all forms of life. Energy via ATP in cells is used to:

- manufacture proteins
- construct new organelles (sub-cellular structures that perform a role within each cell)
- replicate DNA
- synthesize fats and polysaccharides
- pump water through cell membranes
- contract muscles
- conduct nerve impulses

Cellular respiration is the most efficient catabolic process known to exist in nature. Although it occurs in every cell in the body, cellular respiration does not take place simultaneously in the exact same phases throughout all the cells. If the energy produced though cellular respiration was released simultaneously, the body would not be able to process all the energy efficiently, which would result in wasted energy. In addition, the impact of such a large amount of energy being released all at once could overload and damage cells (see "Energy Yields in Cellular Respiration"). As a result, cellular respiration occurs at different stages in the body's various cells, even in cells that are close neighbors or side by side. ATP molecules act like time-release capsules; they release small amounts of energy to fuel various functions within the body at different times.

Overall, two primary processes occur in cellular respiration. The first is the breakdown of glucose into carbon dioxide and hydrogen, known as the *carbon pathway*. The second is the transfer of hydrogen from sugar mole-

Energy Yields in Cellular Respiration

Beginning with a glucose (simple sugar) molecule, each phase of cellular respiration (glycolysis, Krebs cycle, electron transport system) yields various amounts of energies and other important molecules that contribute to creation of cellular energy. In each phase, some energy in the form of ATP is also expended. The following is a summation of the *net* yields and energy gains in each phase of cellular respiration in terms of number of molecules.

Glycolysis
- 2 ATP

- 2 NADH

- 2 Pyruvate

Krebs Cycle
- 2 ATP

- 6 NADH

- 2 $FADH_2$

- 6 CO_2

Electron Transport System
- 32–34 ATP

- 6 H_2O

One glucose molecule after completing the three main stages of cellular respiration yields:

- 36–38 ATP

- 6 CO_2

- 6 H_2O

cules to oxygen, resulting in the creation of water and energy. The entire process of cellular respiration occurs in three primary stages:

1. Glycolysis
2. Krebs cycle (citric acid cycle)
3. Electron transport system

Glycolysis

Glycolysis, which comes from the Greek words *glykos* ("sugar" or "sweet") and *lysis* ("splitting"), is the initial harvester of chemical energy within the body. It occurs in the cell's cytoplasm and converts glucose molecules into molecules of pyruvate, or pyruvic acid. Unlike the other processes in cellular respiration, glycolysis does not require oxygen and is the only metabolic pathway shared by all living organisms. Scientists believe that this biological approach to producing life-giving energy existed before oxygen developed in the Earth's atmosphere. It is the first step in both aerobic (oxygen) and anaerobic (oxygen-free) energy-producing processes (see "Anaerobic Respiration").

Glycolysis is a multi-step process, with each step being catalyzed by a specific enzyme dissolved in the fluid portion of the **cytoplasm** called the *cytosol.* As with all biological processes, energy is needed to begin the process, and two ATP energy molecules initiate the reactions. This initial input of energy is called the *energy investment phase*, and occurs when ATP is used to phosphorylate, or add a phosphate to, the six-carbon glucose molecule. However, the process also yields energy in that further breaking down the six-carbon glucose molecule into two three-carbon pyruvic acid molecules ultimately results in a net gain of ATP molecules, as well as other energy molecules like reduced nicotinamide adenine dinucleotide (NADH). However, glycolysis is extremely inefficient. The entire process captures only about two percent of the energy that is available in glucose for use by

Anaerobic Respiration

When we exercise, our bodies produce more energy and require more oxygen. However, our blood cannot always supply enough of the oxygen via respiration that the cells in our muscles need. Under these circumstances, our muscle cells can respire anaerobically, that is, without oxygen, like some fungi and bacteria are able to do. Anaerobic respiration is also referred to as fermentation. However, cells in the human body can only respire without oxygen for a short period of time.

Like normal aerobic cellular respiration, anaerobic respiration begins with glucose in the cell, but takes place completely in the cell's cytoplasm. Although ATP energy molecules are also produced this way, the process is extremely inefficient compared to aerobic respiration. In the human body, the anaerobic process results in pyruvic acid being turned into the waste product lactic acid, as opposed to entering the mitochondria for further oxidation as it does in aerobic cellular respiration. It is the lactic acid in muscles that makes them stiff and sore after intense aerobic exercise, such as running.

the body. Much more energy is available in the two molecules of pyruvic acid and NADH produced during glycolysis. It is this potential energy that goes on to the next step called the **Krebs cycle.**

The Krebs Cycle (Citric Acid Cycle)

Discovered by Hans Krebs (1900–1981), the Krebs cycle, also known as the citric acid cycle, is a cyclic series of molecular reactions that require oxygen to function. The cycle is mediated by enzymes that help create the molecules for the final harvesting of cellular energy in the third and final phase of the cellular respiration process. The Krebs cycle occurs in the matrix of the **mitochondria**, which are the powerhouses of cells. Although the mitochondrion is the second largest organelle in a cell after the nucleus, some cells may contain thousands of mitochondria from 0.5 to 1 micrometer in diameter. Unlike the energy harvesting process of glycolysis in the cytoplasm, mitochondria are extremely efficient in taking energy from sugar (and other nutrients) and converting it into ATP. In fact, compared to the typical automobile engine, which only harvests about 25 percent of the energy available in gasoline to propel a car, the mitochondrion is more than twice as efficient—it converts 54 percent of the energy available in sugar into ATP.

After glycolysis is completed, the two pyruvate molecules that were formed enter the mitochondria for complete oxidization by a series of reactions mediated by various enzymes. As the pyruvate leaves the cytoplasm and enters a mitochondrion, acetyl coenzyme A (CoA) is produced when an enzyme removes carbon and oxygen molecules from each pyruvate molecule. This step is known as the *transition reaction.*

The Krebs cycle begins as oxygen within the cells is used to completely **oxidize** the acetyl CoA molecules. The process is initiated when each of the acetyl CoA molecules combines with **oxaloacetic acid** to produce a six-carbon citric acid molecule. Further oxidation eventually produces a four-carbon compound and carbon dioxide. The four-carbon compound is ultimately transformed back in oxaloacetic acid so that the cycle can begin again. Because two pyruvate molecules are transferred into the mitochondria for each glucose molecule, the cycle must be completed twice, once for each pyruvate molecule. Each cycle results in one molecule of ATP, two molecules of carbon dioxide, and eight hydrogen molecules. The ATP molecules produced during this cycle can be used as energy. But it is through the cycle's creation of the electron "carrier" coenzyme molecules NADH and reduced flavin adenine dinucleotide ($FADH_2$)—which are created when the **coenzymes** nicotine adenine dinucleotide (NAD) and flavin adenine dinucleotide (FAD) "pick up" the hydrogen molecules—that the abundance of ATP is produced in the next stage of cellular respiration, the **electron transport system.**

Electron Transport System

Overall, the first two processes in cellular respiration, glycolysis and the Krebs cycle, have produced relatively little energy for the body's cells to use. Although both of these processes produce some ATP directly, the energy currency of ATP is created and cashed in for the big payoff during the electron transport system, also known as the electron transport chain. This process takes place across the inner membrane of the mitochondria called the cristae. A chain of electron receptors are embedded in the cristae, which are folded to create numerous inward, parallel, regularly spaced projections or ridges. This design results in an extremely high density of receptors, thus increasing the electron transport chain's efficiency.

The receptors are actually a network of proteins that can carry electrons and transfer them on down a protein chain. The process works like a snowball gaining speed as it rolls down a hill. As the NADH and $FADH_2$ molecules produced during glycolysis and the Kreb's cycle pass down the chain, they release electrons to the first molecule in the chain and so on. Because each successive carrier in the chain is higher in electronegativity (that is, has a higher tendency to attract electrons) than the previous carrier, the electrons are "pulled downhill." During the process, hydrogen protons (H+) or ions from NADH and $FADH_2$ are transferred along a group of closely related protein receptors that include **flavoproteins**, iron-sulfur proteins, quinones, and a group of proteins called **cytochromes**. The cytochrome proteins in the electron transport system will only accept the electron from each hydrogen and not the entire atom. The final cytochrome carrier in the chain transfers the electrons, which by this time have lost all their energy, to oxygen in the matrix to create the hydrogen-oxygen bond of water. This bond is another reason why oxygen is so important to the life of cell. Without it, the molecules in the chain would remain stuck with electrons and ATP would not be produced.

Because of the second law of thermodynamics, the electrons passed down the chain lose some of their energy with every transfer from cytochrome to cytochrome. Some of the energy lost helps to "pump" hydrogen ions out of the mitochondria's matrix into a confined intermembrane space between the mitochondria's inner and outer membranes. This energy for pumping the hydrogen ions is a result of a process called the **oxidation-reduction reaction,** or redox reaction. The reaction results in the molecules within the electron transport system alternately being reduced (gaining an electron) and then oxidized (losing an electron). The entire process establishes a buildup of hydrogen ions resulting in a concentration, or diffusion, gradient—more hydrogen ions are pumped inside the confined space between the mitochondria's membranes than exist in the mitochondria's matrix. As the concentration gradient increases, the ions begin to diffuse back through the membrane into the matrix to equalize the hyrdogen ion gradient.

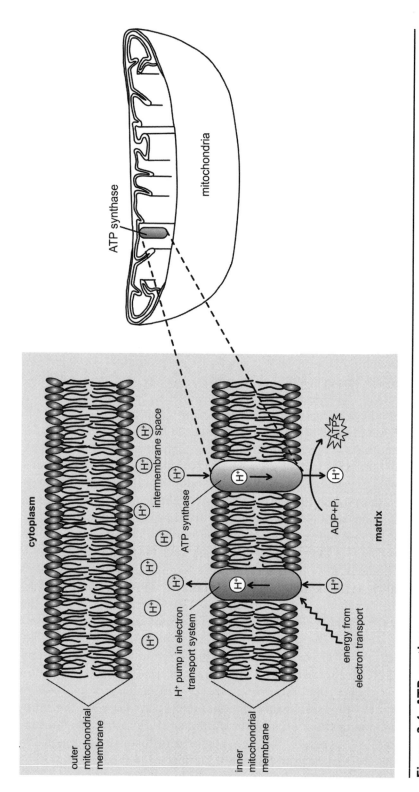

Figure 2.4. ATP synthase.

This diagram illustrates the built up of hydrogen ions into the mitochondria's intermembrane space via electron transport and the eventual transport of these ions back through the membrane where they are used by ATP synthase to make ATP (the major source of energy for cellular reactions) out of ADP and phosphate.

Hydrogen ion diffusion occurs through ATP synthase, an enzyme within the inner membrane of the mitochondrion. ATP synthase uses the potential energy of the proton gradient to synthesize the abundance of ATP out of the adenosine diphosphate (ADP) molecule and phosphate (see Figure 2.4). This process is referred to as *chemiosmosis*. The formation of ATP is an energy storage process, and the energy is released when ATP is converted via the ATPase enzyme back into ADP (adenose bound to two phosphate groups) or to adenosine monophosphate (AMP) (adenose bound to one phosphate group). All of these conversions are known as *ATP phosphorylation*. ADP and the separate phosphates produced by the breakdown are then recycled into cellular respiration for the recreation of ATP. At the same time, the waste products carbon dioxide and water are eliminated via diffusion from the cell into the bloodstream and on through the organismal respiratory process.

3

A Brief History

Ancient ideas about the human body, diseases, and medical treatments were based on a mixture of religious thought, philosophy, and the beginnings of scientific inquiry. Understandably, early thinkers held many misconceptions about the respiratory system. Both Democritus (470–360 BCE) and Aristotle (384–322 BCE), for example, thought that one of the functions of air and breathing was to keep the "soul" in the body. Many ancient Greek philosophers also believed that air created the mind and consciousness, which dwelled in the lungs.

Despite these and other mistaken beliefs, many physicians and seekers of knowledge developed some surprisingly enlightened insights. Democritus, for example, was the first to develop "atomic theory," calling the smallest units of "prime matter" *atoms*. Although Aristotle did not perform anatomical studies on humans, his writings included some basic anatomical descriptions of the respiratory system. He distinguished the windpipe ("air-holder") from the esophagus and had some notion of the larynx and structure of the lungs. He also noted that the lungs received a liberal supply of blood.

This chapter discusses the evolution of knowledge about the respiratory system. It includes a look at early beliefs that became medical dogmas, and how new insights were obtained to ultimately advance knowledge about the respiratory system. In the process, some of the great minds to influence the course of knowledge about the human body are highlighted.

In looking back, it is easy to dismiss our ancestors' ideas and theories about the human body as being naive and misguided. However, they did not even have a compound microscope until almost the 1600s, much less modern, high-tech devices that have provided modern researchers with astounding insights into the biology of the body.

Although many of our ancestors' beliefs about the human body were based on assumptions and proved to be wrong, these attempts to understand how the body works provided the foundation that ultimately ushered in a new world of discovery. This chapter focuses on discoveries up to the mid-1950s; later advancements in understanding the respiratory system in terms of medical treatments are discussed in the chapters about modern medical approaches for treating respiratory disease (see Chapters 6 and 7).

A HISTORY OF THE LUNGS

Establishing an Early Understanding

The study of respiration dates back at least to ancient Egypt and Egyptian healers, who believed that respiration was the most important function in the human body. Because they routinely embalmed the pharaohs, their families, servants, and others for mummification, the Egyptians also had a good idea of what the major organs in the body looked like, including the lungs. But the early Egyptians did not record many of their observations. As a result, they provided little useful knowledge about the body and its major systems for those who followed.

The earliest known anatomical studies of the human body date back to around 4000 BCE, when the people of Mesopotamia in southwest Asia inscribed some observations about the body and its organs on clay tablets. But it was in Alexandria in northern Egypt and Greece that anatomy first became widely accepted as a science that could provide important insights into how the human body functioned.

In the third century BCE, Herophilus of Chalcedon (335–280 BCE) established the first school of anatomy in Alexandria. While the dissection and study of animals was not entirely new, Herophilus was instrumental in helping to educate people to overcome their fears of dissecting the human body. At the time, many believed that tampering with the body might somehow negatively affect the "spirit" in the afterlife, where people would once again assume their human forms.

Herophilus, who taught in Alexandria, is believed to have dissected nearly six hundred human bodies, resulting in a number of anatomical writings. He paid particular attention to the anatomy of the brain and was the first to distinguish clearly between veins and arteries. Herophilus was berated by many for his supposed practice of dissecting live human prisoners, although it is now widely believed that this accusation may not have been true.

One of the earliest "scientific" explanations of respiration and the lungs came from another anatomist named Erasistratus of Chios (c. 80 BCE), who also taught in Alexandria. In his treatise on respiration, Erasistratus said that air was taken in by the lungs and passed to the heart where it entered the blood and changed into a peculiar kind of "pneuma" or spirit, which he called the "vital spirit." It was then sent to various parts of the body by the arteries, where it turned into another type of pneuma called the "animal spirit." Although no knowledge of oxygen or cells was known at the time, Erasistratus's description of how air interacts with lungs and blood is surprisingly similar in its basic overall concept to what we now know about how respiration works.

Much of the work of Erasistratus was further developed by Claudius Galen, a Greek-Roman physician (131–200 CE). Galen used much of Aristotle's theories about the human body to produce his monumental work *De usu partium* (*The Uses of the Parts of the Body*), which would end up becoming a standard medical text until the sixteenth century.

Unlike Erasistratus, Galen did not dissect human cadavers but based most of his writings on animal dissections. He also rejected many of Erasistratus's assumptions. Like Aristotle, Galen believed that the purpose of respiration was to cool the blood by "the substance of the air." This cooling function was to counterbalance the heart, which heated the blood and the body. In essence, Galen believed that inspiration cooled the heart and blood, while expiration removed the hot substances.

Although Galen's writings established many misconceptions about the human body that would last for centuries, he and his forerunners did understand that breathing occurred in the lungs, which they viewed as a type of bellows firing and cooling a furnace. Galen also described the lungs as having "all the properties which make for easy evacuation; for it is very soft and warm and is kept in constant motion." He also correlated lung function with blood movement, noting, "Blood passing through the lungs absorbed from the inhaled air, the quality of heat, which it then carried into the left heart."

Awakening from the Dark Ages

In the centuries following Galen, much of Europe and the west fell into the Dark Ages, a nearly six-century period of cultural stagnation and few scientific advances, including knowledge of the human body. Nevertheless, while Europe languished in ignorance, the Arab world was studying the writings of Hippocrates (c. 460–380 BCE), Galen, and others, building upon their ideas to develop innovative knowledge, medical practice, and medical schools for the times.

In the early part of the eleventh century, for example, Islamic medical philosopher Avicenna (980–1037), who was from what is now called Iran, wrote his *Canon of Medicine*. In his writings, Avicenna noted the unusual moisture of the lungs. He believed the moistness resulted from some type of "nourishment which comes to it." Overall, Avicenna wrote about 240 works that have survived the ages, including forty about the human body and medicine.

In the middle of the thirteenth century, Ibn al-Nafis (1213–1288) made many major contributions to medicine. An eminent Cairo physician and author, he wrote detailed commentaries on earlier medical works, critically evaluating them and then adding his own original thoughts and contributions. His discovery of the blood's circulatory system included the basics of pulmonary circulation, the flow of blood to and from the lungs. It would take European educators three more centuries before they would reach the same level of understanding. He was also the first to correctly describe the constitution of the lungs, including a description of the bronchi.

Although the Dark Ages ended in Europe about 1000 CE and post-classical medical schools began to flourish in the West, knowledge of human anatomy and body functions remained mired in ancient thought. The next significant advances in Europe came during the Renaissance period of enlightenment beginning in the fourteenth century. European thinkers began to revive the ancient Greeks' ideas and use them in their development of new scientific theories, including theories concerning the respiratory system.

Although many of the "experts" in medicine at medieval universities clung to past ideas, some were dedicated to moving forward. Italian anatomist Mondino dei Liucci (c. 1270–1326), who taught at Bologna, performed his own dissections. (Most of his colleagues assigned the distasteful task to their assistants, who seldom had the courage to disagree with their superiors). Based on his own personal observations, dei Liucci published the first practical manual of anatomy in 1316, called *Anothomia*. Nevertheless, he ignored some of his observations, perhaps because of religious and other social pressures, and continued to perpetuate many of the errors of Galen.

A resurging interest in anatomy continued and was urged on by such Renaissance thinkers as Leonardo da Vinci (1452–1519). Working in the latter part of the fifteenth century, Leonardo da Vinci's anatomical drawings demonstrated a keen understanding of the human body. In his notebooks, Leonardo da Vinci explained that "man has within him a pool of blood wherein the lungs as he breathes expand and contract" and that "from the said pool of blood proceed the veins which spread their branches through the human body."

Even though Leonardo da Vinci had no knowledge of oxygen and the role it played in respiration, he did understand that breathing and the lungs helped to discharge waste materials. "From the heart," wrote Leonardo da Vinci, "impurities or 'sooty vapors' are carried back to the lung by way of the pulmonary artery, to be exhaled to the outer air."

Leonardo da Vinci and his contemporaries had initiated a new era of systematic, empirical anatomical studies that would begin to reveal human body systems and how they functioned. Leonardo da Vinci's contemporary, Alessandro Benedetti (1460–1525), built an anatomical theatre where dissections were performed and noted that the "lung changes the breath, as the liver changes the chyle, into food for the vital spirits."

Another founder of modern anatomy was the Italian Andreas Vesalius (1514–1564). Vesalius published two advanced works on anatomy in 1543, including *De Humani Corporis Fabrica* (*On the Fabric of the Human Body*), which was amply illustrated. Vesalius's anatomical studies led him to question much of Galen's findings and ultimately to break with them completely.

In terms of respiration, Vesalius conducted an experiment in which he used a bellows to force air into the lungs of animals immediately after they had stopped breathing. He noted that the heartbeat was restored and the pulse returned. This experiment led to the fundamental conclusion that if the heart can be revived by restoring airflow to the lungs, then obviously the lungs must do more than just "cool" the heart, as

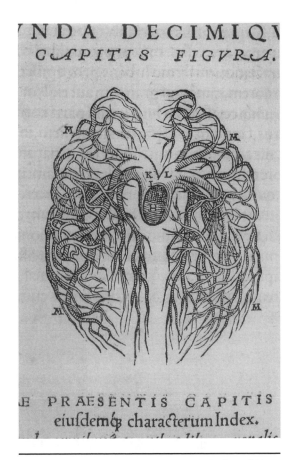

Andreas Vesalius, "Arterial circulation in the lungs." From *De Humani Corporis Fabrica* (*On the Fabric of the Human Body*), 1543. © National Library of Medicine.

was widely believed. Ultimately, Vesalius received a death sentence under the Inquisition (a movement in the Catholic Church that was charged with the eradication of heresies) for robbing graves to acquire bodies for his anatomical studies. The sentence was commuted when he was allowed to make a pilgrimage to the Holy Land.

Realdo Colombo (or Colombus) (c. 1515–1559), who studied with Vesalius, continued his mentor's work and described the circulation of the blood through the lungs and back to the heart. He is also widely credited with introducing the term *circulation* and showed that pumping of the blood occurred when the heart contracted. Throughout this period, anatomical studies further refined knowledge of the lungs, including the existence of the lobes and that the lungs' sponginess facilitated breathing.

Connecting the Lungs, Heart, and Circulation

Another noted sixteenth-century scientist, the Spaniard Michael Servetus (1511–1553), is also credited with making amazing discoveries about pulmonary circulation. Servetus correctly said that blood circulated from the body to the right side of the heart and then through the lungs into the left side of the heart. Without the use of a microscope or any other scientific instrument, Servetus based his observation on "the notable size of the pulmonary artery." He noted that this artery "was not made of such sort or such size, nor does it emit so great a force of pure blood from the heart into the lungs, merely for their nourishment."

Near the end of the sixteenth century, Fabricus, who studied under Colombo and taught in Italy, said that the lungs functioned to prepare air for the heart. But it was one of Fabricus's students, the Englishman William Harvey (1578–1657), who provided the most accurate descriptions in early times of the lungs' significance to life.

Working first on frogs, Harvey noted that so much blood left the heart in a single minute that the amount pumped out far exceeded the total amount of blood in the body. Using mathematical data, he concluded that the blood was not being consumed but circulated throughout the body, directly contradicting Galen, who believed food was converted to blood and then consumed by the body. Harvey also disagreed with his own contemporaries who felt that the lungs were responsible for moving blood around the body.

Continuing his studies of circulation, Harvey ultimately detailed pulmonary circulation in the lungs. By 1615, he had completed a description of the closed circulatory system and the mechanics of blood flow through the lungs and heart, including identifying that the heart sends blood to the lungs to be "refreshed." Harvey waited until 1628 to publish his findings, partly because so much of what he wrote rejected the "sacred" teachings of Galen and, unless proven, could have ended Harvey's career.

A physician, Harvey also demonstrated that fetal circulation bypasses the lungs because the lungs were collapsed and inactive. He stated that healthy lungs were extremely important to proper functioning and even noted that exercise was an important way to maintain lung health.

THE DISCOVERY OF OXYGEN

Defining the "Vital Spirit"

Despite the many advances being made in anatomy and the growing knowledge that the lungs played an extremely important role in the body's ability to function, Harvey and many others remained deeply committed to traditional thought. For example, Harvey and his contemporaries maintained the long-held view that certain organs are responsible for people's emotions and character traits. In the case of lungs, they believed that people who had hot lungs were bold and those with cold lungs were timid.

By the early sixteenth century, however, some were questioning these theories. Shortly before his death, Miguel Servet (1511–1553) wrote that inhaling air into the lungs could have other functions. He noted that "the Vital spirit" of life is a "mixture of inspired air with the most subtle portion of the blood." Servet and others suspected that something in the air was being absorbed into the lungs and blood to provide the basis for life, but they still had no idea exactly what "it" was.

More than a century after Servet, many were still questioning the function of the lungs. In 1667, Englishman Robert Hooke (1635–1703) presented his famous experiment with a dog to the Royal Society. Using a bellows to inflate and deflate the lungs, Hooke was able to keep the animal alive for more than hour. Hooke at first believed that it was the inflation and deflation of the lungs that kept the dog alive. In further experiments he provided the lungs with air without having them inflate and deflate, and the dog still remained alive, showing that it was not the movement of the lungs themselves that was important. This turn of events convinced him that it was the renewal of air that was important in keeping the animal alive. Hooke hypothesized that respiration could be an action where "something in the air" is exchanged into the blood system and then carried to other parts of the body.

At almost the same time that Hooke was doing his work, John Mayow (1640–1679) was conducting experiments on respiration and the air. His observation on the power of saltpeter to set fire to organic substances led him to speculate that the air contained what he called "vital particles" that were essential to life. These "vital particles" would eventually prove to be oxygen. In the process of his work, Mayow also made important observations on ventilatory movements, explaining how the intercostal muscles create the vacuum that leads to filling the lungs with air. He also described the role of the diaphragm with considerable accuracy.

Peering Deep Inside the System

By the late 1500s, a technological breakthrough opened up a whole new world to scientists. Two Dutch spectacle makers, Hans and Zaccharias Janssen, started experimenting with lenses, leading the father and son team to invent the compound microscope (a microscope that uses two lenses).

The Janssens focused their microscope by sliding two tubes, each with a lens attached. The lens in the eyepiece tube was biconvex, meaning that it bulged outwards on both sides. The lens of the far end tube was planoconvex, that is, flat on one side and bulging outwards on the other side. The advanced microscope had from three to nine times power of magnification.

In the early 1600s, Galileo Galilei (1564–1642) improved on the discovery with a focusing device. The Dutchman Antony van Leeuwenhoek (1632–1723), however, came to be known as "the father of microscopy" for his improvements. Refining the method for grinding and polishing small lenses, Leeuwenhoek developed magnifications up to 250 diameters. As a result, he was the first to see and describe bacteria and the circulation of blood as corpuscles in capillaries.

Marcello Malpighi (1628–1694), who was the physician to Pope Innocent XII, used the microscope to discover the smaller blood vessels and the blood corpuscles. In a letter to his friend, the mathematician Giovanni Morelli, Malpighi also wrote the first description of the alveoli. He described them as "an almost infinite number of orbicular sinused vesicles, just as in a honeycomb we see [alveoli] formed by wax spread out into walls." In the course of studying the alveoli, Malpighi recognized a membranous wall through which gases were exchanged. He observed the capillary network throughout the lungs and discovered similar capillaries in the other organs as well. He also showed how circulating blood passes from arteries to veins.

In 1669, Richard Lower (1632–1691) made the important observation that blood turns from a dark to a bright red color as it passes through the lungs. He correctly attributed this change to the exposure of blood to air. Although this observation had been made earlier, it was originally attributed to the friction of blood moving through the lungs rather than to the effects of air. Lower went on to make many observations about pulmonary circulation.

Advancing Physiology through Chemistry

The next great advance in understanding respiration came through the field of chemistry. In the eighteenth century, chemists began isolating gases in the air and their different roles in the respiratory process. During his research on magnesium carbonate in the early 1750s, Scottish physician Joseph Black (1728–1799), who laid the foundation for modern chemistry, performed an experiment in which air was exhaled through limewater. The experiment produced a gas that was both distinct from air and yet a part of atmospheric air. Black called the gas "fixed air"; it is now known as carbon dioxide.

In the 1770s, the chemists Joseph Priestly (1733–1804) and Carl Wilhem Scheele (1742–1786) discovered the existence of oxygen. Scheele of Sweden found what he called "fire air" through several experiments in combustion, including heating silver carbonate or mercuric carbonate. He published his findings in *Chemische Abhandlung von der Luft und dem Feuer* (*Chemical Treatise on Air and Fire*) in 1777. Scheele demonstrated that air, when freed from "aerial acid" (carbon dioxide) and water vapor, consisted of two gases: "fire air" (oxygen) that supports combustion, and "foul air" (nitrogen) that does not. Scheele went on to discover chemical elements (chlorine, manganese, molybdenum, tungsten, and barium) and several molecules, including citric acid and hydrogen fluoride. Although Scheele actually made his discoveries first, he published his findings after Priestly because he was waiting for a preface to be written for his treatise. As a result, it is Priestly's name that is most commonly associated with the discovery of oxygen (see "The Origins of Oxygen").

The clergyman-scientist Priestly isolated the same substance discovered by Scheele by heating mercury using a magnifying glass. Due to high densities, mercury will not absorb gases easily like water. Priestly would heat

The Origins of Oxygen

Oxygen is the vital component of the earth's atmosphere, the elixir that makes almost all life possible, from microorganisms to fish to humans. However, evidence shows that more than 2.5 billion years ago there was little free-floating oxygen in the atmosphere. Instead, the atmosphere was made up mostly of nitrogen, carbon dioxide, and water.

The earth's atmosphere changed radically around two billion years ago as it became saturated with oxygen. In a sense, it was the planet's first great pollution crisis. Why a crisis? Because oxygen can kill many bacteria and other simple organisms. As a result, organisms had to develop biochemical methods for rendering oxygen harmless, such as oxidative respiration within cells to produce energy.

Where did the oxygen come from? The most popular theory is that a type of bacteria called *cyanobacteria* proliferated. This photosynthetic organism produces oxygen as a byproduct and appeared around 3.5 billion years ago. However, cyanobacteria only became widespread during the Proterozoic Era, around 2.5 billion years ago. Overall, scientists believe it took another 300 to 500 billion years for the cyanobacteria to produce enough oxygen in the atmosphere to support complex life forms.

Another more recent theory is that geological processes, such as volcanic gases, initiated the rise in oxygen levels. According to some scientists, the Earth's mantle (the region below the crust and above the core), where volcanic gases originate, contained an abundance of "reducing" components like iron silicates. These components react with oxygen and extract most of the oxygen produced by photosynthesis. For a variety of reasons, volcanic gases became more oxidized over time, and their ability to "mop up," or reduce, atmospheric oxygen when they were released into the air decreased. As a result, the net photosynthetic production of oxygen was greater than the volcanic sink, resulting in an oxygen-rich atmosphere.

Several other theories have also been proposed regarding the reason for high levels of oxygen forming in the atmosphere. But regardless of the origins of oxygen, its presence changed conditions on earth forever. Without it, the only life on earth would be tiny microbes.

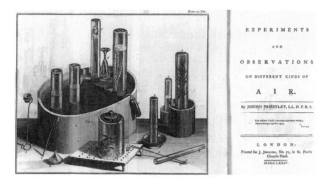

Frontispiece and title page of Joseph Priestly, *Experiments and Observations on Different Kinds of Air*, London, 1774–1777. Courtesy of the Library of Congress.

various substances in a tube with mercury essentially using a magnifying glass and, with an apparatus he developed, collect the gases over mercury. The first gas he found was nitrous oxide, which quickly became known as laughing gas for its effect on people and later developed into a surgical anesthetic.

When heated to near its boiling point (675°F–356.7°C), mercury oxidizes in air, forming mercuric oxide. Priestly decided to burn a piece of mercuric oxide in the test chamber and found that when he placed a smoldering wood sliver into a sampling of the gas it produced, the wood burst into flame. He also found that a candle would burn brightly in the gas, while the other gases he had tested extinguished the flame. Priestly called the gas "dephlogisticated air." (The term came from the word "phlogiston," the name given to a supposed material substance without color, odor, taste, or weight that scientists thought was released when something burned.) Priestly did not know it, but he had discovered that at 932°F (500°C), mercuric oxide decomposes into mercury and oxygen.

Priestly's experiments were wide-ranging and included identifying the gases involved in plant respiration, thus unifying the knowledge gained in chemistry and biology. He was the first to identify photosynthesis. Priestly also accidentally invented soda water. While experimenting with brewery gas, he noticed that a gas had drifted to the ground around the brewery vat. He deduced that this gas was "heavier" than normal air. When Priestly dissolved the heavy gas in water, he liked the tangy taste it left behind. The gas was carbon dioxide, and Priestly's invention of soda water helped earn him election to the French Academy of Sciences in 1772 and a medal from the Royal Society in 1773.

Discovering the Chemistry of Respiration

The discoveries of Black, Scheele, Priestly, and others laid the groundwork for a French chemist to make extraordinary discoveries about respiration. Antoine Laurent Lavoisier (1743–1794) was the first to uncover the intricacies of respiration chemistry. His work also led to him to accurately formulate the complete process of pulmonary gas exchange.

Lavoisier was a brilliant chemist studying combustion when Priestly traveled to Paris to discuss his experiments with him. Lavoisier almost immediately realized what Priestly had uncovered: air contained two substances and not two different kinds of air—one with phlogiston and one without—as Priestly and others believed.

Lavoisier conducted experiments that invalidated the phlogiston theory. He went on to call the part of air that supported combustion "oxygen" (from the Greek words meaning "acid-producing" because he mistakenly believed that oxygen was a necessary component of all acids). He also called the part of air that did not support combustion "azote" (from the Greek for "no life"), which is now known as nitrogen.

But Lavoisier did much more than discover the existence of oxygen. In 1780, he conducted experiments with guinea pigs in closed chambers showing that they consumed oxygen and produced carbon dioxide. Lavoisier described the respiratory process this way: "Eminently respirable air ('oxygine') that enters the lung, leaves it in the form of chalky aeriform acids (carbon dioxide) in almost equal volume. . . . Respiration acts only on the

Antoine Laurent Lavoisier is known as the father of modern chemistry and conducted groundbreaking research into respiration and oxygen. © National Library of Medicine.

portion of pure air that is eminently respirable . . . the excess, its mephitic portion (nitrogen) is a purely passive medium which enters and leaves the lung without alteration."

Lavoisier also worked with Armand Seguin (1767–1835) to study metabolism and observed that oxygen consumption, pulse rate, and the respiratory rate increased during work. Around 1789, the duo made the fundamental physiological connection between external respiration and internal combustion, or energy production. They showed that respiration provided the necessary ingredients for heat (energy) production liberated by metabolic processes.

Lavoisier and Sequin believed that the body's heat production occurred only in the lungs when hydrogen and carbon were oxygenated by the intake of air. It took nearly fifty years before German physiologist Heinrich Gustave Magnus (1802–1870) analyzed blood and decided that "combustion" must occur throughout the body and not just in the lungs. His famous experiments also revealed that arterial and venous blood contained both carbon dioxide and oxygen. And he discovered that the arteries contained a higher ratio of oxygen to carbon dioxide as compared to the veins.

Gaining New Insights into the Respiratory Process

Although the discovery of oxygen was by no means the last piece of the respiratory puzzle, its discovery, coupled with the discovery of carbon dioxide, served as the centerpiece around which new research focused. Henry Cavendish (1731–1810) had also previously identified hydrogen as a substance distinct from other flammable gases in 1766. For the first time, chemists and physiologists truly had a strong foundation for studying the chemical reactions in the body that consumed oxygen and produced carbon dioxide.

The beginning of the nineteenth century initiated the modern era of understanding the respiratory process. Although he is known primarily for debunking the theory of spontaneous generation of life in microorganisms, Lazzaro Spallanzani (1729–1799) also conducted research into the process of respiration, focusing primarily on amphibians. In 1803, four years after his death, his important findings on respiration were published. He was the first to prove that tissues (or cells) use oxygen and give off carbon dioxide.

At about the same time, Thomas Garnett made a presentation to the London Royal Society in which he hypothesized that oxygen was absorbed by the iron in red blood cells. This hypothesis was carried on by Justus von Liebig (1803–1873), who found an iron compound in red blood cells and no iron in any other tissues. He then inferred that oxygen arriving in the lungs must be absorbed by the blood as it gave off carbon dioxide.

Other major discoveries in terms of human oxygen usage included the first crystallization of hemoglobin in 1862 by German physician Ernst Felix Hoppe-Seyler (1825–1895). The physician not only gave hemoglobin its name, but he also clarified its role in red blood cells, including its conversion into oxyhemoglobin. In the late 1860s, while working in the lab of Hoppe-Seyler, Johann Friedrich Miescher (1844–1896) discovered that the amount of carbon dioxide in the blood affects the respiratory rate in humans. The physiologists Karl Ludwig (1816–1895) and Eduard Pfluger (1829–1910) also studied the gas exchange process in the blood. In the early 1870s, they showed that oxygen uptake and biological oxidation necessary for the processes of life occurred in tissues and discovered the existence of respiratory enzymes in tissues.

MAJOR DISCOVERIES OF THE TWENTIETH CENTURY

Uncovering the Details of Gas Transport

In the first decade of the twentieth century, two physiologists, Christian Bohr (1855–1911) and John S. Haldane (1860–1936) set out to discover why blood took up and released much higher levels of oxygen than were expected based solely on the concentration of oxygen in the lungs, or oxygen pressure. The two found separate effects that showed that the oxygen-binding capacity of the blood is determined not only by oxygen pressure but also by pH, or the concentration of hydrogen and carbon dioxide. In the process, they unraveled the basic route that oxygen and carbon dioxide take in the gas exchange process.

In 1904, Bohr, who was the father of the famous physicist Neils Bohr, discovered that the affinity of hemoglobin for oxygen depends on acidity or pH levels in tissues. Carbon dioxide builds up in the tissues as a waste product of cellular respiration. It then enters the blood stream in the capillaries and affects the unloading point of hemoglobin. When carbon dioxide is present, the hemoglobin will unload more readily as its affinity for hanging on to its oxygen decreases. In other words, Bohr had found that an increase of carbon dioxide in the blood resulted in the dissociation of oxygen from hemoglobin and other respiratory compounds.

The shift in the affinity of hemoglobin for oxygen became known as the Bohr effect. Of special interest was the fact that what happens takes place where it is most desirable for hemoglobin to unload its oxygen—in the capillaries surrounding tissues. Because of the discovery of the Bohr effect, scientists could begin to better understand how the intake of oxygen is facilitated in the lungs while oxygen is more easily given off by the blood surrounding tissues. In effect, Bohr managed to explain how the blood transports and releases oxygen to tissues.

At about the same time as Bohr, Haldane was studying the exchange of gases during respiration, primarily in connection with his interest in the health hazards of coal mining and deep-sea diving. Haldane discovered that when blood is saturated with oxygen, the amount of carbon dioxide carried by the blood is reduced. He also found that, upon reaching the capillaries surrounding the alveoli, oxygen influx into the capillaries from the lungs increased the release rate of carbon dioxide into the alveoli.

This became known as the Haldane effect, which says that oxygen removal from hemoglobin increases hemoglobin's affinity for carbon dioxide and vice versa. As a result, Haldane further clarified how blood transports and releases carbon dioxide into the lungs. Essentially, carbon dioxide hitches a ride with hemoglobin at the site of the tissues, where it is produced as a waste product, and then emptied out into the lungs as the alveoli provide an influx of oxygen. Scientists then understood how carbon dioxide could be carried away from the tissues without a large increase in the partial pressure of carbon dioxide, or a large fall in pH or acidity, as the blood goes through the systemic capillaries.

In 1905, Haldane also published his theory that breathing is controlled by the effect of the concentration of carbon dioxide in arterial blood on the respiratory center of the brain. He noted that his experiments indicated "clearly that under normal conditions the regulation of the lung-ventilation depends on the pressure of carbon dioxide in the alveolar air. Even a very slight rise or fall in the alveolar carbon dioxide pressure causes a great increase or diminution in the lung-ventilation."

Explaining Oxygen Transfer

August Krogh (1874–1949), who studied in the laboratory of Christian Bohr and once served as Bohr's laboratory assistant, made several important discoveries concerning respiration. In 1903, Krogh's experiments with frogs showed the differences between tissue, specifically the skin, and lung respiration. He found that skin respiration was relatively constant but that great variations occurred in lung respiration, which was controlled by the **autonomic nervous system** and the vagus nerve. He surmised that respiration rates varied according to the need for oxygen.

Krogh pursued his interest in the fundamental question of how the gas exchange process occurred between the lungs and the blood. Many scientists believed in the secretion theory, that is, that the lung served as a type of gland that secreted oxygen from the alveoli to the blood. To better study this process, Krogh invented the microtonometer—an instrument that measures gas pressure in fluids. He had also designed earlier equipment while working with Bohr that enabled his mentor to measure the oxygen-binding capacity of blood and thus discover the Bohr effect.

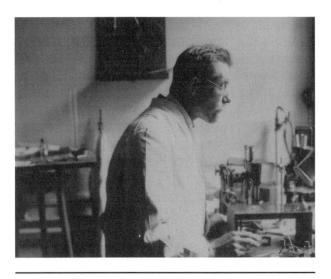

August Krogh discovered that oxygen passed into the blood from the lungs through passive diffusion. © National Library of Medicine.

Using the microtonometer, Krogh was able to precisely determine the gas content of blood and quickly found that there was no difference between the oxygen content of blood and the alveoli. As a result, he suspected that the passage of oxygen into the blood was a passive diffusion process. He also disproved the secretion theory by showing that there is no secretion of oxygen into the air sacs of fish.

Krogh revealed his results in 1907 to Bohr, who supported the secretion theory. Krogh's study clearly showed that it was simple diffusion and not any active secretion process that caused the exchange of oxygen and carbon dioxide between the blood and the lungs. The discovery shattered Bohr's authority on the subject, and Krogh's former mentor never spoke to him again.

Krogh struggled with publishing his findings because he respected Bohr and because he wanted to be absolutely certain he was correct. He knew that contradicting a renowned expert in the field could mean the end of his career if he was wrong. In 1910, he finally published his results, noting: "The absorption of oxygen and the elimination of carbon dioxide in the lungs take place by diffusion and by diffusion alone. There is no trustworthy evidence of any regulation of this process on the part of the organism."

Also interested in the mechanisms that allowed blood capillaries to supply oxygen to muscle cells and to remove carbon dioxide in large volumes as required by exercise, Krogh discovered that capillaries in the organs and tissues close at rest but open during exercise or other activities, when the need for oxygen carried by the blood increases. The conventional view had been that these capillaries always remained opened and that the speed of blood flow increased during exertion.

Krogh correctly questioned this belief, arguing that faster blood flow through capillaries did not increase the potential for more oxygen supply because the time for diffusion would decrease. Instead, Krogh measured the oxygen content in capillaries and muscle fibers and showed that a relatively small number of capillaries are open when humans are at rest. He also showed that capillaries are opening and closing all the time, and that during work more capillaries are open. The oxygen supply, Krogh revealed, increased because so many capillaries opened during exertion and each capillary could hold even more blood.

Krogh's major findings opened a window through which scientists could view an array of mechanisms that enabled the human body to meet its oxygen needs. Krogh himself went on to provide new insights into such issues as the binding of gasses in blood, gas transport by blood flow, and the exchange of oxygen and carbon dioxide in the tissues. In 1920, Krogh won the Nobel Prize for physiology or medicine for his research into human respiration (see "Nobel Prize Winners").

Nobel Prize Winners

Over the years, many scientists have won the Nobel Prize in physiology or medicine and in chemistry for research and discoveries related to the respiratory system. Some of these discoveries have focused on the respiratory system itself, such as the work of Corneille Jean Francois Heymans on the regulation of respiration. Others received recognition for research related to the cellular workings that are part of cellular (internal) respiration, which involves the production of energy through cells incorporating the gas exchange process between the tissues and blood circulating from the lungs. The following is a list of Nobel Prize winners who have helped to advance knowledge concerning the respiratory system and process. If needed, a further explanation of a prize-winner's work in relation to the respiratory process is supplied in parentheses.

Nobel Prize in Chemistry

1978   Peter Mitchell "for his contribution to the understanding of biological energy transfer through the formulation of the chemiosmotic theory." (This theory sheds light on the mechanisms by which electron transfer is coupled to ATP synthesis in oxidative phosphorylation; ATP is the "energy currency" of the cell produced during cellular respiration.)

1997   Paul D. Boyer and John E. Walker "for their elucidation of the enzymatic mechanism underlying the synthesis of [ATP]."

Nobel Prize in Physiology or Medicine

1905   Robert Koch "for his investigations and discoveries in relation to tuberculosis."

1920   August Krogh "for his discovery of the capillary motor regulating mechanism." (This mechanism is related to the transport of materials via the blood stream, including oxygen.)

1922   Otto Fritz Meyerhof "for his discovery of the fixed relationship between the consumption of oxygen and the metabolism of lactic acid in the muscle."

1931   Otto Heinrich Warburg "for his discovery of the nature and mode of action of the respiratory enzyme."

1937   Albert von Szent-Györgyi "for his discoveries concerning the biological combustion processes." (These processes are related to oxidation in cells and the liberation of energy.)

1938   Corneille Jean Francois Heymans "for the discovery of the role played by the sinus and aortic mechanisms in the regulation of respiration."

1945   Sir Alexander Fleming, Sir Ernst Boris Chain, and Lord Howard Walter Florey "for the discovery of penicillin and its curative effect in various infectious diseases." (Penicillin has been used to treat respiratory infections.)

1947	Carl Ferdinand Cori and Gerty Theresa Cori "for their discovery of the course of the catalytic conversion of glycogen." (Glycogen is a source of glucose, which supplies cells with an oxidizable energy source in cellular respiration.)
1952	Selman Abraham Waksman "for his discovery of streptomycin, the first antibiotic effective against tuberculosis."
1953	Sir Hans Adolf Krebs "for his discovery of the citric acid cycle." (The cycle is also known as the Krebs cycle.)
1955	Axel Hugo Theodor Theorell "for his discoveries concerning the nature and mode of action of oxidation enzymes." (A one-time coworker with Nobel Prize Winner Warburg, Theorell further clarified the oxidation process within cells.)
1992	Edmond H. Fischer and Edwin G. Krebs "for their discoveries concerning the reversible protein phosphorylation as a biological regulatory mechanism." (Its many regulatory functions includes serving as an important mechanism responsible for the regulation of glucose from glycogen in cellular respiration.)

Finding the Yellow Enzyme of Cellular Respiration

Among the most accomplished biochemists of all time, Otto Heinrich Warburg (1883–1970) was interested in the chemical process of oxidation and the vital processes of the human body. The production of cellular energy had long been correlated with combustion, or burning things in atmospheric oxygen at high temperatures, thus producing a form of energy. Warburg and others knew that something happened in cells that utilized small substrates (substances acted upon by enzymes) from ingested food to be converted into usable energy by living cells. But they had no idea what it could be.

Warburg decided to focus on finding the substance that initiated the energy-producing process, but no ordinary chemical methods were available for Warburg to isolate the respiratory catalyst for cellular respiration. Warburg turned to indirect methods. Knowing that metals possess the power to initiate or accelerate various reactions, including combustion, Warburg assumed that intracellular "combustion" could be due to a metal catalyst. This thought was bolstered by Warburg's discovery that small amounts of cyanide inhibited cell oxidation. Because cyanide can combine with heavy metals like iron to form stable complexes, Warburg theorized that a catalyst important to oxidation must contain a heavy metal.

To help with his research, Warburg developed a manometer (a pressure gauge for comparing pressures of a gas) for monitoring cell respiration and measuring the rate of oxygen production in living cells. Warburg then developed a technique for getting extremely thin slices of living tissue and keeping them alive in a medium filled with specific nutrients. He placed the tissue under different conditions and monitored the tissues changes as they consumed oxygen for respiration.

Warburg found that what he called the "respiratory ferment" (his name for the catalyst of cellular respiration) had a tendency to combine with iron, making him believe that the catalyst is due to iron. He also isolated a yellow flavoprotein, or enzyme, from yeast, which was shown to catalyze the oxidation of another vitamin cofactor (now known as NADPH) in cellular respiration, using oxygen as the final substrate to create energy. Warburg's yellow enzyme proved to be the catalyst for the oxidation-reduction action that is necessary for cellular respiration.

Warburg had shown that cellular respiration was enzymatic in nature and eventually went on to discover the mechanism by which iron was involved in the cellular use of oxygen, demonstrating that many properties of the respiratory enzyme are due to the iron molecules it contains. In 1931, Warburg received the Nobel Prize in physiology or medicine "for his discovery of the nature and mode of action of the respiratory enzyme." Ultimately his work identified the importance of cytochromes as proteins that promote chemical reactions in living cells and how they affect cellular respiration.

Warburg also conducted many other experiments showing how plants assimilate carbon dioxide in photosynthesis and that carbon monoxide inhibited respiration. He revealed the metabolism of tumors and made the classical observation that glycolysis is enhanced in tumors, observing that malignant cell growth required much less oxygen than that of normal cells.

Revealing the Regulation of Breathing

Scientists had known since the 1880s that blood pressure changes were linked to changes in the rate and depth of breathing. The most widely accepted theory as to why this occurred was that the brain's respiratory center in the medulla was affected by either the direct action of blood pressure or the rate of blood flow in cerebral circulation.

Corneille Jean Francois Heymans (1892–1968) won the 1938 Nobel prize in physiology or medicine for disproving this theory and revealing new facts about how breathing is regulated in humans. Heymans initially set out to disprove another observation by Heinrich E. Hering. Hering had noted that a reflex action in the carotid artery (the two major arteries on the left and right side of the neck) influenced the heartbeat. For his counterexperiment, Heymans developed the "isolated head" technique.

Heymans deattached an anesthetized dog's head from its body with only the vagus aortic nerves intact and no connection to the brain. The dog also shared circulation with a second anesthetized dog to keep its body functioning. During the course of his experiment, Heymans discovered that respiratory response to changes in blood pressure ceased when the aortic nerves were severed. He had irrefutable proof that the medulla and the brain did not play a role in this particular reflex action. The reflex mechanism's sole sensory pathway were the aortic nerves.

Heymans eventually found that pressure-sensitive areas called *pressore-ceptors* had a reflexive reaction in response to blood pressure rates. He discovered the pressoreceptors in the wall of the carotid sinus, which is a slight enlargement in the carotid artery where the artery divides into the external and internal carotids.

In addition to uncovering this previously unknown way for the body to regulate respiration, Heymans later discovered chemoreceptors in the cardio-aortic and carotid sinuses areas that responded to changes in the chemical composition of blood. His other contributions included major findings in the physiology of cerebral circulation and the physiology of blood circulation during exercise.

Finding the Source of Cellular Energy

In the early 1930s, scientists knew little about the intermediary stages by which sugar from foodstuff is oxidized in cells to create energy. Many of the essential processes connected with the metabolism of the cell to create energy had been uncovered. But the overall process behind cellular energy production eluded scientists. It was Hans Krebs (1900–1981), a former student of Otto Heinrich Warburg, who would piece the puzzle together and present a clear picture of the essential principle of how cells use oxygen as part of cellular respiration to release energy.

Attracted to the problem of intermediary pathways in cell oxidation, Krebs used a series of elegant experiments to work out the cyclic nature of the energy producing reactions in cells, first within the mitochondria of eukaryotes (organisms with one or more cells containing visible nuclei and organelles) and later within the cytoplasm in prokaryotes (cellular organisms, such as bacteria, that do not have a distinct nucleus). Krebs theorized that certain acids with two **carbocyclic** acid groups (for example, oxaloacetic acid and fumaric acid) should operate in a cyclic fashion to oxidize glucose and amino acids within the cells.

Using the breast muscles of pigeons because the tissue was metabolically active and maintained its oxidative capacity in solution, Krebs noted that only certain organic acids were readily oxidized by muscle and that they turned out to be substrates of the tricarboxylic acid enzymes. He then discovered the missing link that allowed the known reactions to create a cyclic sequence involving citric acid. In summary, the Krebs cycle regenerates oxaloacetate acid through a series of intermediate compounds while liberating carbon dioxide and electrons that are immediately used to form high-energy phosphate bonds in the form of ATP, which creates the cell's energy reserves.

The Krebs cycle is extremely important because it explains two simultaneous processes. First it showed how degradation reactions yield energy from foodstuff. The cycle also explained the building-up processes in the cells that use up energy. In essence, the cycle proved to be the essential and common metabolic pathway in the breakdown of carbohydrates, fats, and proteins from foodstuff into carbon dioxide and water to generate energy within the cell.

The Krebs cycle is also known as the *tricarboxylic acid cycle*. Later it also came to be called the *citric acid cycle* after researchers established that citric acid was the first substrate formed in a vital reaction. Initially, many of Krebs's colleagues criticized him, and the prestigious journal *Nature* rejected his manuscript. But he soon began to gather supporters for his findings. For his discovery, Krebs received the 1953 Nobel Prize in physiology or medicine.

4

Respiratory Problems and Diseases

Few people have gone through life without experiencing an acute (short but severe) upper respiratory infection, more commonly known as the common cold. Not only are colds the most prevalent infectious disease known, respiratory ailments and diseases as a group are more common than any other medical problem in humans.

While the familiar runny nose, sneezing, and sore throat that usually accompany a cold are relatively benign, other respiratory conditions are much more serious. For example, pneumonia can be life threatening, especially in children, the elderly, and people weak from other ailments or diseases. Another respiratory disease, lung cancer, has become the leading cause of cancer death in the United States for both men and women.

Numerous factors can adversely affect respiratory functioning, from genetic influences and medical problems during infancy to overall health as we grow older. Even psychological well-being may impact respiratory health. For example, some cases of bronchial asthma have been linked to anxiety, and a sudden anxiety attack can lead to **hyperventilation**. Nevertheless, environmental factors are by far the most common cause of respiratory ailments and diseases.

Because we breathe in approximately 16,000 quarts (over 15,000 liters) of air each day, our respiratory system is exposed to a continuous barrage of substances in the air that can affect the system's functioning, from bacteria and viruses to pollutants caused by industry and automobiles. According to some estimates, a person inhales and ingests approximately 10,000 microorganisms per day. Although the respiratory system is designed to protect our bodies against this environmental onslaught, it is not always successful, especially in cases of overexposure to pollutants. Cigarette smoking, for example, is directly responsible for the overwhelming majority of lung cancer and emphysema cases.

THE RESPIRATORY SYSTEM'S DEFENSE MECHANISMS

The respiratory system has several features that help protect it from the possible harmful effects of environmental particles and **pathogens** (viruses, bacteria, etc.) that can enter the system when we breathe. In the upper respiratory tract, the **mucociliary** (mucus and ciliary) lining of the nasal cavity is the respiratory system's first line of defense. Composed of tiny hairs lining the nose, this defense mechanism filters out the particles inhaled from the environment. The second line of defense is the mucus that lines the turbinate bones (scrolled spongy bones of the nasal passages) in the sinuses and collects particles that get past the nose. These defense mechanisms together trap larger particles from 5 to 10 micrometers in diameter.

As the air we breathe passes through the nose and nasal cavity, it enters the pharynx, where many particles also stick to the mucus on the back of the throat and tonsils. These captured particles can then be eliminated via coughing and sneezing. In addition, the adenoids and tonsils in the back of the throat help trap pathogens for elimination. These lymphoid tissues (tissue from the lymphatic system) also play an important role in developing an **immune system** response, such as the production of **antibodies** to fight off germs.

The lower portion of the respiratory tract also has ciliated cells and mucus-secreting cells that cover it with a layer of mucus. These features work together with the mucus-trapping particles and pathogens, which are then driven upwards by the sweeping ciliary action to the back of the throat where they can be expelled.

Most of the upper respiratory tract surfaces (including the nasal and oral passages, the pharynx, and the trachea) are colonized by a variety of naturally occurring organisms called *flora*. These organisms (primarily of the staphylococcus group) can help to combat infections and maintain a healthy respiratory system by preventing infectious microorganisms or pathogens from getting a foothold. This phenomenon is known as *colonization resistance* or *inhibition*, and occurs because the normal flora compete for space and nutrients in the body. Some flora also produce toxins that are harmful to other pathogenic microorganisms. In rare instances, normal flora can help cause disease if outside factors cause them to become pathogenic or they are introduced into normally sterile sites in the body.

Despite these defense mechanisms, pathogens and particles from 2 to 0.2 micrometers often make their way to the lungs and the alveoli. For example,

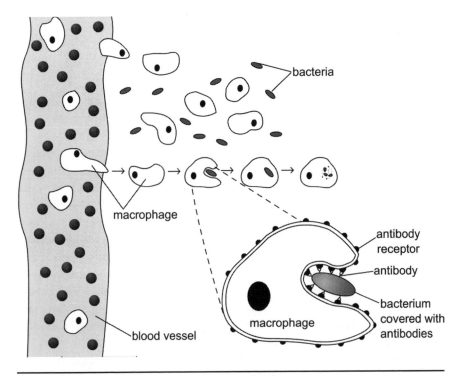

Figure 4.1. Macrophages.
This diagram illustrates how a macrophage, which means "big eater," engulfs, or ingests, bacteria and other microbes. It does this by surrounding the particle to be eaten; then the macrophage's membrane flows together and the particle ends up inside.

most bacteria and all viruses are 2 micrometers or smaller. The alveoli, however, also have defense mechanisms to protect against microscopic invaders. In the case of the lungs, these mechanisms are primarily cellular in nature. For example, alveolar **macrophages** are a type of leukocyte that ingest and destroy invading organisms as part of the immune system's response to infection (see Figure 4.1). The fluid lining the alveoli contains many components, such as surfactant, phospholipids, and other unidentified agents, that may be important in activating alveolar macrophages. Lymphoid tissue associated with the lungs also plays a role in defending against infections by initiating immune responses. For example, immune system cells, such as B and T cells, represent a local immune response to fight off infections by producing antibodies or activating macrophages.

RESPIRATORY SYSTEM PROBLEMS AND DISEASES

While the respiratory system's defense mechanisms are largely effective in battling infections, environmental pathogens can still cause problems

Viruses and Bacteria Are Not the Same

Because bacteria and viruses cause many familiar diseases, especially in the respiratory system, people often get them confused or think that they are the same type of microbes. In fact, viruses are as different from bacteria as plants are from animals.

Bacteria have a rigid cell wall and a rubbery cell membrane that surround the cytoplasm inside the cell. Within the cytoplasm is all the genetic information that a bacterium needs to grow and to duplicate or reproduce, such as deoxyribonucleic acid (DNA), ribonucleic acid (RNA), and ribosomes. A bacterium also has flagella so that it can move (see Figure 4.2).

Despite the minute size of bacteria, viruses are much smaller. Viruses are surrounded by a spiky layer called the *envelope* and a protein coat. They also have a core of genetic material, either in the form of DNA or RNA (see Figure 4.3). Unlike bacteria, viruses do not have all the materials needed to reproduce on their own. As a result, they invade cells, either by attaching to a cell and injecting their genes or by being enveloped by the cell. Once inside the cell, they harness the host cell's machinery to reproduce. Viruses eventually multiply and cause the cell to burst, releasing more of the virus to invade other cells.

Some scientists do not define viruses as "living" but rather view them as packets of floating information that only come to life when they encounter and invade a suitable host. Antibiotics fight bacterial infections by killing bacteria or preventing them from growing. They are useless for viral infections because they cannot kill something that is not alive. Although some bacterial infections, like bacterial tonsillitis, are quite common, viruses are responsible for the majority of respiratory infections.

when a sufficient "dose" of an infectious agent is inhaled. For example, during the cold and flu seasons, a larger quantity of viruses and bacteria are alive and circulating in the air (see "Viruses and Bacteria Are Not the Same"). If they enter the respiratory tract and gain a foothold so that they overcome the body's defense mechanisms and colonize respiratory tract surfaces, the individual will "catch" a cold or the flu.

Not all respiratory system problems are caused by infections or environmental assault, such as pollution and cigarette smoke. For example, **cystic fibrosis** is a genetic disease that can affect the respiratory system by producing an overabundance of thick mucus that can eventually close the respiratory system airways.

The respiratory problems and diseases discussed in this chapter focus on general respiratory ailments that primarily affect the upper respiratory tract:

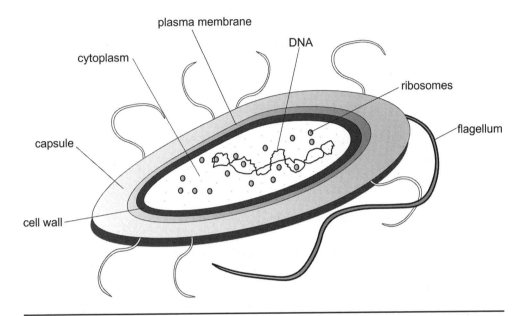

Figure 4.2. Bacterium.

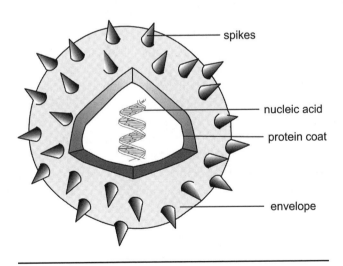

Figure 4.3. Virus.

- Epistaxis (bloody nose)

- Laryngitis

- Pharyngitis

- Rhinitis (also known as hay fever)

- Sinusitis (sinus infection)

- Sleep apnea

- Strep throat

- Tonsillitis

- Upper respiratory infections (common cold)

The next chapter focuses on the usually far more serious lung disorders and diseases.

Epistaxis (Bloody Nose)

Epistaxis, or a bloody nose, is usually little more than a nuisance and not serious in nature. There are two primary types of nosebleeds. Anterior epistaxis is when bleeding occurs in the front part of the nose and is a relatively benign and temporary condition. In posterior epistaxis, the bleeding occurs in the back of the nose and may indicate a more serious problem.

The vast majority of nosebleeds occur when minute blood vessels lining the inside of the nose break and bleed, usually when the relatively thin nasal mucosa overlying the blood vessels dries, scabs, and falls off. This often happens when sneezing, blowing, coughing, or picking has irritated the nose. Extremely dry air, common during the winter or in very arid climates, can also dry out the nasal membrane and cause nose bleeds. The second most common cause of epistaxis is trauma to the nose. However, in some cases, epistaxis may be caused by a more serious underlying condition, such as hypertension, coagulopathies (diseases affecting blood coagulation), renal (kidney) failure, cirrhosis, cancer or tumors of the nose or nasal septum, and hereditary hemorrhagic telangiectasia (also called Rendu-Osler-Weber disease, in which capillary vessels are abnormally dilated).

Treatment: In most cases not involving an underlying medical problem, epistaxis can be stopped by several simple approaches, including applying ice to the forehead, pressure to the upper lip, or pinching of the nostrils against the septum. If the bleeding continues, nasal packing can be used (for example, wadding up a tissue into the nose). Physicians may prescribe medications for more serious problems, such as a topical vasoconstrictor, which narrows the blood vessels, like oxymetazoline hydrochloride or epinephrin, which can be followed by **cauterization** of the bleeding site with silver nitrate or electric cautery. Individuals who are susceptible to nose bleeds can also take various precautions to prevent epistaxis, like not lifting heavy objects or forcefully blowing the nose, keeping the nose moist with a spray salt water solution, and using a humidifier to add moisture to the air.

Laryngitis

Inflammation of the larynx, or laryngitis, occurs when the vocal cords in the larynx become swollen or scarred. The vocal cords include a central bundle of muscles and various layers of connective tissue covered with mucosa. Any change in these layers can cause hoarseness, low voice, and scratchy throat.

Laryngitis results most often from a viral or bacterial upper respiratory infection. It may also occur when the voice is strained or misused. For example, many people such as auctioneers and singers can develop laryngitis, and over-enthusiastic yelling can strain the larynx. Cigarette smoke, dust, or other airborne pollutants are also causes of laryngitis. In rare cases, laryngitis may be linked to a more serious illness such as growths on the vocal chords, allergies, and Horner syndrome, which leads to laryngeal paralysis.

For the most part, laryngitis is temporary in nature, especially if the source is viral, and will heal within a week or so. Most chronic (long-term or frequently recurring) cases are due to continuous voice strain or exposure to environmental pollutants. However, severe cases can result in fever and other respiratory problems. In children, for example, laryngitis can more easily lead to respiratory obstruction because their airways are much more narrow. This problem most often occurs when laryngitis is combined with conditions known as **epiglottitis** and **laryngotracheobronchitis**, also known as "the croup."

Treatment: Treatment for laryngitis varies somewhat according to the cause. For example, if it results from a bacterial infection, a doctor may prescribe antibiotics. However, most cases result from viral infections and strain, and resting the voice by talking as little as possible is the primary way to reduce vocal cord inflammation. Air humidifiers, taking hot showers, and drinking plenty of fluids may also relieve hoarseness and other symptoms. Over-the-counter medications, such as decongestants and analgesics, may also provide relief for some of the symptoms when they are due to an upper respiratory infection. Doctors often recommend that people suffering from laryngitis avoid smoking and drinking alcohol.

Pharyngitis

Pharyngitis is the medical term for a sore throat, or inflammation of the pharynx. It is estimated that viruses cause forty to sixty percent of pharyngitis cases, and only fifteen percent are caused by a bacterial infection known as strep throat (see later section in this chapter). Pharyngitis often accompanies an upper respiratory infection, or the common cold. Few complications are associated with a sore throat due to a viral infection. However, severe pharyngitis due to other ailments such as mononucleosis can lead to airway obstruction and other problems.

Treatment: Doctors only prescribe antibiotics if pharyngitis is caused by a bacterial infection, because antibiotics are not effective against viruses. Pharyngitis associated with a viral infection usually goes away without medication. Otherwise, treatment primarily involves easing the symptoms by gargling warm salt water and taking an over-the-counter pain relief medication. In cases where the tonsils are chronically infected, a surgical procedure to remove the tonsils may be required (see section on tonsillitis later in this chapter).

Rhinitis

Rhinitis is an inflammation of the mucous membrane lining the nose and the sinuses. The primary symptoms include a runny nose, sneezing, and nasal congestion, and may include itchy nose, throat, and eyes. Although rhinitis usually is not a serious problem in terms of impacting a person's overall health, it is an irritating and uncomfortable ailment. However, many cases of allergic rhinitis, particularly when it occurs in children, can lead to various other complications, including chronic otitis media (ear infection), rhinosinusitis, conjunctivitis (eye infection), and sinusitis. Although there are many types of rhinitis, they are broken down into two main categories: allergic rhinitis (more commonly known as hay fever) and nonallergic rhinitis.

ALLERGIC RHINITIS (HAY FEVER)

Although **allergies** come in many forms, allergic rhinitis is the most common allergic disorder known to medicine. It affects approximately 40 million people in the United States and results in an estimated 10 million lost days of school or work each year. For the most part, it occurs in people who are sensitive to airborne irritants called *allergens*, including pollen, dust, animal dander, fungus, molds, and grasses. However, allergic rhinitis can also result from allergic reactions to other substances.

In allergic rhinitis, allergens trigger the release of antibodies against the various allergens. These antibodies attach themselves to mast cells, a type of **leukocyte** or white blood cell that contains various **inflammatory mediators**, including **histamines, prostaglandins**, and **leukotrienes**. The mediators cause inflammation and fluid production in the linings of the nasal passages and sinuses.

Treatment: Physicians use many approaches to treat allergic rhinitis, including oral and inhaled medications, **immunotherapy**, and allergy shots. Perhaps the best and most effective approach for treating allergic rhinitis is to avoid the causes or allergens as much as possible. For example, some people suffer from seasonal allergies that occur at certain times of the year, mostly in the spring and fall when high concentrations of pollens are in the air. Hot, dry, windy days are also more likely to have increased amounts of pollen in the air than cool, damp, rainy days when dampened pollen is "washed" from the air to the ground. At these times and under certain conditions, allergy sufferers should avoid prolonged or vigorous activity outside and keep the doors and windows closed in their homes and cars. For ongoing problems due to allergins inside the home, many people must avoid having pets and make an extra effort to control dust and molds in the home to help relieve their symptoms.

Many pharmaceutical (prescription and nonprescription) and medical approaches are available to provide relieve from allergic rhinitis. Although antihistamines are perhaps the most broadly used pharmaceutical approach to treating allergic rhinitis, they are not always effective because the mediators causing the symptoms could be prostaglandins or leukotrienes and not histamines. The following are the primary approaches to treating allergic rhinitis:

- Oral antihistamines inhibit the action of histamine on the nasal passages and eyes.

- Nasal decongestants decrease nasal tissue swelling and stuffiness.

- Nose sprays (nonprescription) block the allergic reaction on the mast cell by preventing or reducing the release of histamine and other chemical substances which cause the allergic symptoms.

- Corticosteroids (nasal and oral) are hormones that can help reduce the inflammation, mucus, and swelling in the nose and sinuses.

- Eye drops contain an antihistamine decongestant to relieve allergic symptoms of red, itchy, watery eyes.

- Saline (salt water) nose drops (nonprescription) are sometimes helpful in relieving nasal symptoms.

- Immunotherapy usually involves a series of injections, or "allergy shots," containing the allergens believed to be triggering allergy symptoms, with the goal of reducing the individual's sensitivity to specific allergens, thus reducing the irritating symptoms.

NONALLERGIC RHINITIS

Although the symptoms of allergic rhinitis are identical to those of the nonallergic form, nonallergic rhinitis does not result from hypersensitivity to allergens. The many types of nonallergic rhinitis include:

- Infectious rhinitis is caused by an upper respiratory viral or bacterial infection.

- Eosinophilic rhinitis accounts for 20 percent of all rhinitis cases and produces elevated eosinophil (a type of leukocyte) counts in nasal samples, or smears. Scientists have not completely determined its causes.

- Occupational rhinitis is usually the result of inhaling irritants such as grains, wood dusts, and chemicals in the workplace.

- Hormonal rhinitis is due to hormonal imbalances resulting from pregnancy, hypothyroid states, puberty, oral contraceptive use, or conjugated estrogen use.

- Drug-induced rhinitis is due to the use of certain medications or drugs, such as reserpine, beta-blockers, aspirin, nonsteroidal anti-inflammatory drugs (NSAIDs), inhaled cocaine, oral contraceptives, and prolonged use of nasal sprays.

- Gustatory rhinitis follows consumption of certain foods, including hot and spicy foods.

- Vasomotor rhinitis is believed to result from a disturbance in the regulation of the autonomic nervous system, which regulates individual organ function and homeostasis (maintaining stable physiological conditions). It results in vasodilation (widening of the blood vessels) and edema (excess accumulation of fluid) of the nasal vasculature.

- Mechanical obstruction rhinitis is associated with a deviated nasal septum or enlarged adenoids.

Treatment: Treating nonallergic rhinitis depends largely on the source of the problem. For example, surgery may be required in the case of a deviated septum, and antibiotics are used to treat bacterial infections. For the most part however, the approach to treating and preventing nonallergic rhinitis is very similar to allergic rhinitis, including avoiding the causes and using medications such as antihistamines, decongestants, saline solutions, and intranasal steroids for patients with symptoms that are not easily controlled with other approaches.

Sinusitis (Sinus Infection)

Sinusitis, or the common sinus infection, shares a strong resemblance to rhinitis in that it includes inflammation of the membranes that line the nasal passages. However, most sinusitis cases are due to an upper respiratory infection and include inflammation of the paranasal sinuses. In acute cases of sinusitis, the most common causes are viral infections. Fungal infections can cause sinusitis as well, especially in people who have allergies to fungi. In fact, people who suffer from allergies and allergic rhinitis often develop chronic sinusitis. Growths called polyps may block the sinus passages and result in sinusitis in some individuals.

Approximately 37 million Americans each year suffer at least one episode of sinusitis. The most common symptoms in sinusitis are a feeling of throbbing pressure and pain in the sinus areas as well as tenderness in the upper face caused by swelling in the nasal passages. This swelling results in air and mucus getting trapped behind the sinuses, resulting in nasal congestion and blockage and sometimes by a mild fever. A runny nose, headache, and fatigue are also common symptoms.

Treatment: Sinusitis is treated differently depending on the cause. If a bacterial infection is the source of the problem, the doctor may prescribe antibiotics. Decongestants are often used to reduce congestion. Some people take pain relievers if severe pain and headaches are present. Chronic sinusitis is often much more difficult to treat and persists even with the use of antibiotics and decongestants. In some cases of long-term sinusitis, oral and nasal steroids may be prescribed, but these medications can have severe side effects and are usually used as a last resort. In extreme cases, sinus surgery may be performed to enlarge the natural opening to the sinuses. People suffering from sinusitis may use a steam vaporizer, over-the-counter saline nasal sprays, and gentle heat applied over the face, such as a warm washcloth, to ease the symptoms.

Sleep Apnea

Apnea comes from the Greek word meaning "without breath." Sleep apnea is characterized by a brief stoppage in breathing during sleep that can last from ten seconds to over a minute and occurs as many as hundreds of times during the night. First described in 1965, sleep apnea is a common disorder that affects more than 12 million Americans.

The most common type of sleep apnea is obstructive sleep apnea, which occurs when the upper airways become blocked. This blockage usually results from the soft palate at the base of the tongue and the uvula (the small fleshy tissue that hangs from the center of the back of the throat) collapsing or sagging and partially or completely blocking the airway during sleep (see Figure 4.4). Overweight people often have this type of sleep apnea because of excess tissue in the airway. The other type of sleep apnea is *central sleep apnea*; it occurs because the brain fails to send signals to the breathing muscles (chest muscles and diaphragm) to make them work.

People with sleep apnea begin breathing again because the brain senses lowered levels of oxygen and increased levels of carbon dioxide in the blood and alerts the body to arousal, that is, to wake up and start breathing again. As a result, people with sleep apnea do not get the same prolonged, restorative sleep as other people. People with sleep apnea will often snore or make choking sounds during the night and repeatedly wake up. This cycle may repeat during the day because the person is often tired from lack of sleep. However, not everybody with sleep apnea snores, especially those with the rare form of central sleep apnea.

Sleep apnea has many serious implications and consequences. One of the most common is increased sleepiness during the day, which hinders concentration and performance in such areas as work and driving a car. Untreated sleep apnea can also cause headaches and serious physical

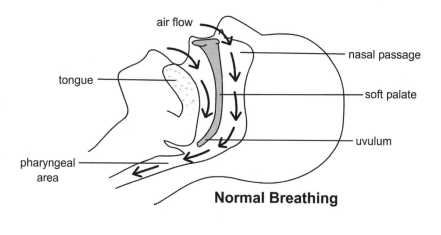

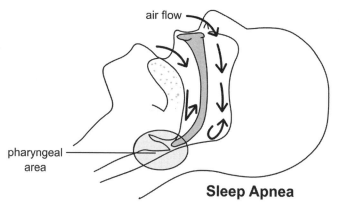

Figure 4.4. Obstructive sleep apnea.
This diagram shows how airflow obstruction is caused within the pharynx, or pharyngeal area, in obstructive sleep apnea. Collapse begins when the base of the tongue abuts the pharyngeal wall and soft palate.

problems, including high blood pressure resulting from the heart pumping harder to make up for oxygen drops in the blood. Eventually, sleep apnea can lead to other cardiovascular diseases and increase risk for heart attack and stroke. A person's memory and sexual functioning may also be affected.

Although sleep apnea can occur in anyone, it is most common in men. Overweight people and people who snore often are more likely to develop sleep apnea. Other risk factors include high blood pressure and physical abnormalities in the upper airways. Smoking and alcohol also increase the risk. Because sleep apnea sometimes occurs repeatedly in families, some cases may have a genetic cause.

Treatment: Because many people who snore and wake up often in the night do not have sleep apnea, physicians test for sleep apnea with a polysomnography test that looks at body functions during sleep, such as electrical activity of the brain and eye movement. No medication has been found to treat sleep apnea. Often, treatments focus on lifestyle changes, such as avoiding alcohol and cigarettes and losing weight. Studies have shown that a ten percent weight loss greatly reduces the number of apneic events in many patients. Sometimes sleep apnea only occurs when the person sleeps on their back, and adjusting to sleeping on their sides often helps.

Nasal continuous positive airway pressure, or CPAP, is the most effective and common treatment for obstructive sleep apnea patients. It involves wearing a mask over the nose during sleep while the CPAP machine pushes air through the nasal passages at a pressure adjusted to prevent the throat from collapsing (see photo). Sometimes dental appliances are used to reposition the jaw and tongue. Surgical approaches include a procedure called *uvulopalatopharyngoplasty*, or UPPP, to remove excess tissue at the back of the throat. A tracheotomy may be per-

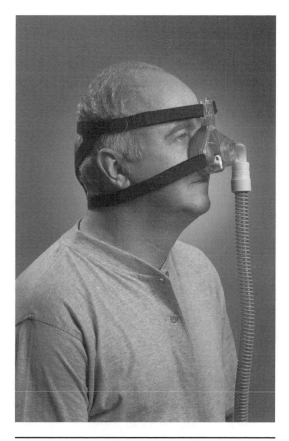

Sleep apnea patient wearing the "Respironics Profile™ Lite Nasal Gel Mask." © Respironics, Inc. Murrysville, PA.

formed in a very few, severe, life-threatening cases. It involves making a small hole in the windpipe and inserting a tube. This tube remains closed during the day but is opened at night so air flows directly into the lungs.

Strep Throat

During the cold and flu seasons, many people develop sore throats, or pharyngitis, primarily due to viral infections. Strep throat, however, is a sore throat due to an infection by a type of *Streptococcus* bacteria. In addition to a severely sore throat, strep throat symptoms include fever and headache. Children may also have stomach pains, nausea, and/or vomiting. Interestingly, most cases of strep throat do not include a stuffy nose or cough like sore throats that are caused by viral infections.

It is difficult to tell whether or not a sore throat is strep throat just by the symptoms alone. As a result, doctors diagnose strep throat by taking a **throat**

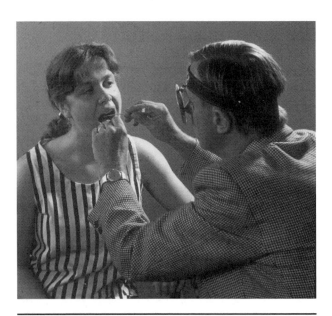

A doctor takes a throat culture from a patient. © SPL/Custom Medical Stock Photo.

culture (see photo) to determine if the *Streptococcus* bacteria are the culprits. It is important to diagnose and treat strep throat because it can lead to other problems, including scarlet fever and impetigo, a skin infection that is more likely to occur in younger children. Other rare complications include rheumatic fever, which can lead to heart disease and arthritis, and nephritis, an inflammation of the kidneys that results in bleeding into the kidney and urine. Strep throat could also lead to sinusitis, ear infections, and pneumonia. Approximately 5 percent of children who develop a sore throat and fever have strep throat caused by Group A streptococci.

Treatment: If it is determined that the sore throat is due to *Streptococcus* bacteria, the doctor will prescribe an antibiotic to treat it, usually penicillin. According to the *Journal of the American Medical Association*, it is important to complete the antibiotic regimen prescribed by the doctor to ensure that all of the bacteria have been killed and to avoid more serious complications. A number of things can be done at home to relieve the symptoms of strep throat, including gargling with warm salt water and drinking hot liquids, such as tea with lemon and honey.

Tonsillitis

Tonsillitis is an infection of the tonsils, which are the **lymph glands** that lie in the back of the mouth and top of the throat. The tonsils help to prevent infection by filtering out bacteria and other microorganisms. It is also believed that tonsils play a role in helping the body's immune system fight disease. For example, the tonsils contain natural killer cells, a specific type of disease-killing cell. Nevertheless, the tonsils themselves can become infected if too much of a bacteria or virus infiltrates the glands. Often, tonsillitis is accompanied by pharyngitis and/or strep throat. In addition to a sore throat, people with tonsillitis often have difficulty swallowing. They may also experience headaches, fevers and chills, and loss or change of voice. Tonsillitis is a common ailment, especially in children.

If left untreated, tonsillitis can lead to peritonsillar abscess, a condition in which pus is formed due to tissue death and the breakdown of white blood cells that normally fight infections. If too much swelling occurs due to the abscess, the airway can be blocked and breathing can become difficult. In rare cases, tonsil infections can spread into the neck and chest and cause life-threatening complications. Complications also include dehydration due to difficulty swallowing fluids, rheumatic fever, and kidney failure.

Treatment: Treatment depends on whether or not the infection is due to a bacteria or virus. Doctors prescribe antibiotics for bacterial infections. In most cases of viral infection, sore throat symptoms are eased by drinking warm or very cold fluids, gargling with warm salt water, or using throat lozenges available in most drug stores. In cases of chronic or ongoing tonsillitis, surgical removal of the tonsils, called a *tonsillectomy*, may be required. However, a tonsillectomy is performed primarily only when other treatments have failed and the problem is chronic.

Upper Respiratory Infections (Common Cold)

According to the National Institute of Allergy and Infectious Diseases, as many as 1 billion upper respiratory infections may occur each year in the United States. More widely known as the common cold, these infections can be caused by more than 200 different viruses. The primary viruses responsible for the common cold are called *rhinoviruses* (from the Greek word *rhin*, for nose), which cause an estimated 25 to 35 percent of all colds. (Rhinoviruses may be the main cause of colds because they grow best at 91.4°F or 33°C, which is the temperature of human nasal mucosa.) Other viruses that can causes colds include the myxoviruses (such as the influenza and parainfluenza viruses), coronaviruses, and adenoviruses. Bacterial agents cause approximately 10 percent of colds.

Viruses are transmitted or spread from person to person in several ways. Studies have shown that cold viruses reach their highest concentration in the nasal secretions three to four days after infection, which means this is when the infected person is most contagious and likely to pass on the virus. One common way of catching a cold, or most viral or bacterial infections, is to touch almost anything that an infected person has also touched, sneezed on, or coughed on, from a doorknob to a telephone to their hands. (Some viruses, such as the human immunodeficiency virus, or HIV, cannot be caught in this manner.) After touching the surface, the virus can be transmitted to the body when the person then touches their nose or eyes, which have ducts that drain into the nasal cavity. Inhaling droplets in the air resulting from someone sneezing or coughing close to you is also a common way to catch a cold (see "Common Myths About the Common Cold").

Viruses cause colds when they penetrate the nasal mucosa, after which they enter cells lining the nasal region and the pharynx. Rhinoviruses, for example, bind to a molecule much like a docking system in a space station.

Common Myths about the Common Cold

MYTH: Cold weather and catching a chill increases the chances of catching a cold.

FACT: Research has shown that neither of these factors increases a person's susceptibility to a cold. For the most part, colds are more prevalent in the cooler weather of late fall, winter, and early spring months because people are indoors more, where they are in closer contact with people and where little fresh air circulates and viruses are more likely to spread.

MYTH: Chicken soup can help cure a cold.

FACT: This home remedy actually has no effect on "curing" the cold or eliminating the virus from the body. However, eating chicken soup can help relieve the symptoms of a dry scratchy throat and, like other hot drinks, may help open nasal passages and relieve congestion. Chicken is also full of protein, and the broth contains electrolytes for nourishment to help fight off the cold. Some studies have found preliminary evidence that chicken soup may help inhibit the production of a type of white blood cells called *neutrophils* that promote cold symptoms such as runny nose and a cough.

MYTH: Flu shots can cause colds.

FACT: Since vaccines are made from an inactivated virus, there is no association between flu shots and colds. However, in some cases, people may be allergic to a vaccine and develop certain cold-like symptoms such as fever and muscle aches.

MYTH: Vitamin C and herbal remedies prevent or reduce the length of colds.

FACT: No indisputable scientific evidence exists that vitamin C or other vitamin and herbal remedies actually prevent colds. Some studies have shown that vitamin C and zinc lozenges may help reduce a cold's duration by a day or so, but the evidence is not conclusive. In most cases, megadoses of the vitamin were given, which can have risky side effects, including diarrhea, an increased intake of sodium (if the vitamin comes in the form sodium ascorbate), and increased and excessive ingestion of iron with damage to tissues.

MYTH: Drinking milk when you have a cold can increase nasal mucus.

FACT: Although milk and mucus look similar, it is not turned into nasal mucus.

MYTH: Your immune system is weakened if you catch a cold.

FACT: Although a weakened immune system may play a factor in some people developing a cold, anyone exposed to enough of the virus can catch a cold. In one study, approximately 95 percent of adults with healthy immune systems who were exposed to the virus via droplets in the nose caught a cold.

MYTH: Sneezing and blowing the nose to get rid of excessive mucus help a person get over a cold quicker.

FACT: Sneezing and nasal secretions do help to remove dust and pollen from the nose, but do not eliminate cold viruses because the virus multiplies inside the nasal cells where it is safe. Although blowing the nose to eliminate built-up mucus is necessary, it actually sends nasal secretions into the sinus cavity, where the virus and other factors can produce more inflammation leading to secondary bacterial infection. Sneezing and coughing also help spread the virus to others.

MYTH: Kissing is a good way to catch a cold from someone.

FACT: Contrary to popular belief, mouth saliva does not contain large amounts of a common cold virus. Nevertheless, cold viruses can be transmitted via kissing; but it is not a very effective or common mode of transference.

Specifically, they contain depressions on their protein shell, sometimes referred to as "canyons," that fit onto surface protein receptors on the nasal cells known as the intercellular adhesion molecules, or ICAMs. This provides the portal for the virus to enter into the cell and begin replicating. It ultimately reproduces thousands of copies of itself leading to cell disruption and release into the nose, where the infection is further spread to nearby nasal epithelial cells.

The primary symptoms of the common cold, including runny and/or stuffed up nose, sneezing, coughing, and scratchy or sore throat, are thought to occur when the body produces an immune response to fight off the viral invasion. White blood cells that are part of the immune system are sent to the site of infection, where they release a variety of chemicals, such as **kinins**. These natural "disinfectants" actually cause the nasal membranes to swell and inflame, proteins and fluid to leak from capillaries and lymph vessels, and mucus production to increase. Other cold symptoms include itchy eyes, fever (usually mild), fatigue, and headache.

Susceptibility: Numerous factors may make a person or a group of persons more susceptible to catching a cold. For example, colds occur more often in children because they usually are in close contact in schools and day-care centers with other children who may have colds. In addition, most children do not always practice good personal hygiene, such as washing their hands and covering their mouths and nose when they cough, which can help prevent the spread of the cold viruses. A very young child's immune system also has not yet fully developed, especially in terms of resistance to infections. For many viral infectious diseases, a person may develop resistance once they have come in contact with a viral infection. In essence, the immune system remembers the infection and builds defense mechanisms to stop it from reoccurring in the future. As a result, colds become rarer and rarer in people as they grow older.

Nevertheless, adults still get colds because so many different viruses can cause them. The wide variety of viruses is also why it may be impossible to ever develop a **vaccine** to prevent the common cold. When people reach a very old age, their immune systems may also begin to be less effective. Another major factor in susceptibility to colds is increased exposure to other people with the virus in crowded areas, such as office workplaces. Other factors that increase susceptibility include allergic disorders affecting the nasal passages or pharynx, menstrual cycles in women, and exposure to cigarette smoke and other environmental pollutants.

Some medical conditions may also increase susceptibility, including AIDS, some cancers, and certain genetic disorders, such as sickle cell disease, cystic fibrosis, and Kartagener's syndrome (which causes cilia to malfunction). Numerous research studies have also related high psychological stress levels to increased susceptibility. Although it is not clear why this connection exists, researchers believe that stress affects the immune system, leading to an increased release of hormones and other substances that may decrease resistance to infections.

Treatment: Despite the many advances made in modern medicine, no one has yet to find a way to actually treat the common cold or prevent it. For example, antibiotics are ineffective because they only work against bacteria, which causes very few of the common cold cases. However, antibiotics may be used when a secondary bacterial infection develops in the sinuses, ears, lungs, or throat after a cold provided bacteria a better opportunity to grow in these areas by increased inflammation, secretions, and swelling.

The only way to treat a cold is to treat the symptoms. In addition to bed rest, doctors recommend drinking plenty of fluids, gargling with warm salt water, using petroleum jelly for a sore nose, and taking aspirin or acetaminophen to help relieve headaches and fevers. Other nonprescription medications, such as decongestants and cough suppressants, may help relieve some of the symptoms but have no effect on the duration of the illness, which is self-limiting and usually lasts a few days to a week or so.

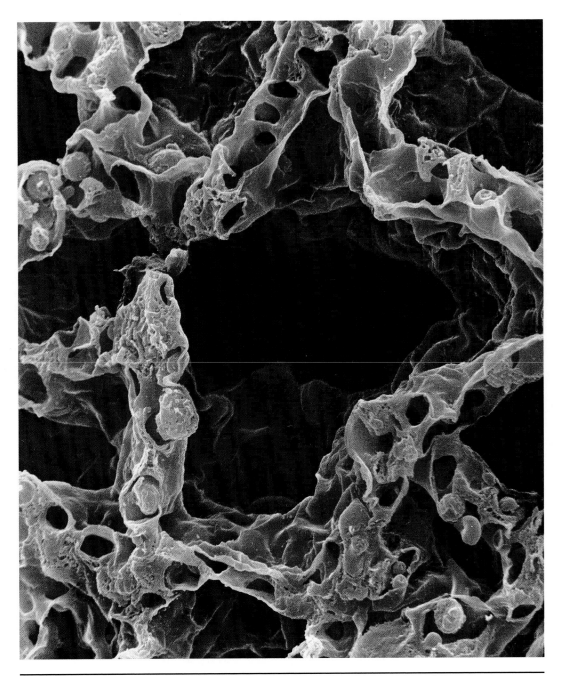

Alveolus in the lung. © Dr. David Phillips/Visuals Unlimited.

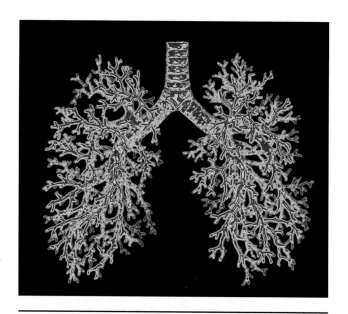

Computer-enhanced image of a resin cast of the airways in the lungs. The trachea (windpipe, top center) divides into two bronchi, which divide further into small bronchioles. The bronchioles terminate in alveoli (not seen), grape-like clusters of air sacs surrounded by blood vessels. Here the blood takes up oxygen and releases carbon dioxide to be exhaled. © Alfred Pasieka/Photo Researchers.

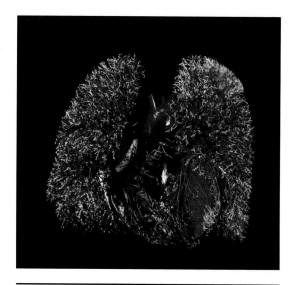

Resin cast of the blood vessels of the human heart and lung, viewed from the front. © Martin Dohrn/Royal College of Surgeons/Photo Researchers.

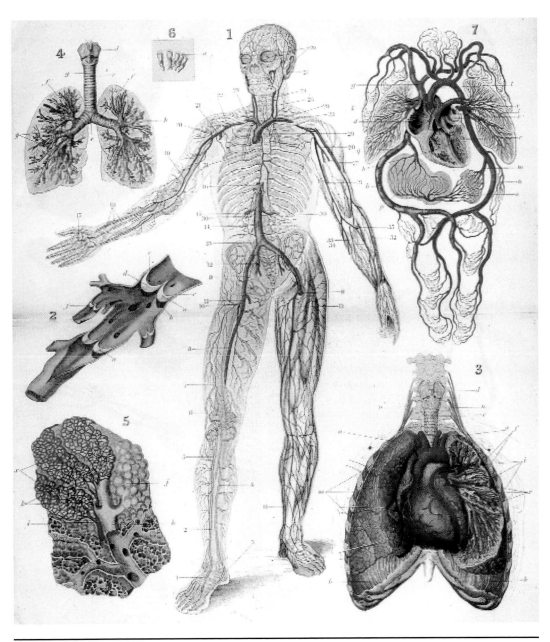

"The veins and lungs." From William Turner and John Goodsir, *Atlas of Human Anatomy and Physiology*, Edinburgh, 1857. © National Library of Medicine.

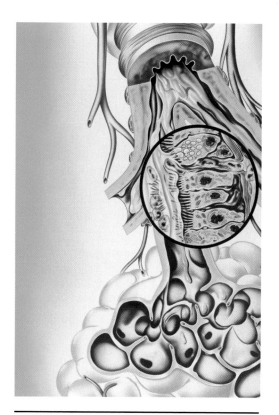

Cut-away illustration of the respiratory system showing the effects of asthma. The trachea is at top and the alveoli are at bottom. At upper right the cutaway bronchioles reveal their excessive sputum, or phlegm (green). In asthma, bronchiole narrowing results in inflammation and shortness of breath. In response, mucus-producing goblet cells (one seen magnified at upper center) produce too much sputum. This coats the cilia of the bronchiole lining (epithelium). © John Bavosi/Photo Researchers.

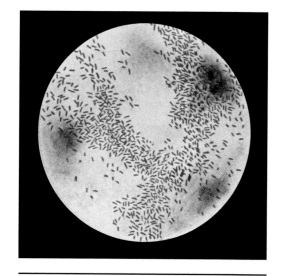

Photomicrograph using a Gram stain technique of the flu bacteria *Haemophilus influenzae*, which was first identified during the flu pandemic of 1918. © Centers for Disease Control and Prevention.

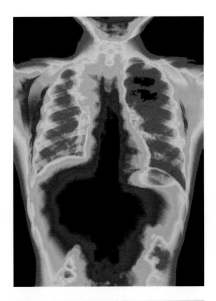

Lobar pneumonia. Colored chest x-ray showing lobar pneumonia in the upper lobe of a 7-year-old girl's lung. The affected area appears pale blue in the top of the red lung at upper left. © S. Camazine/Photo Researchers.

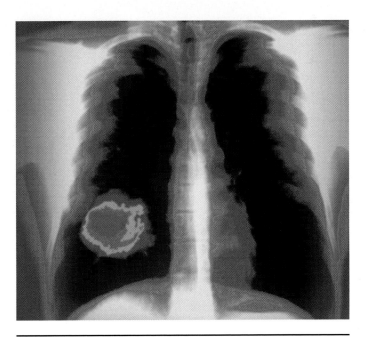

X-ray showing cancer of the right lung (in yellow). © ISM/Phototake.

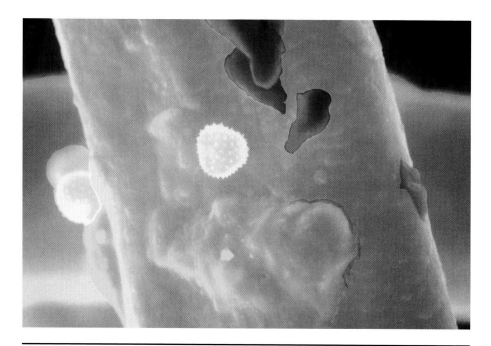

Ragweed pollen attached to nasal hair mucus. © JL Carson/Custom Medical Stock Photo.

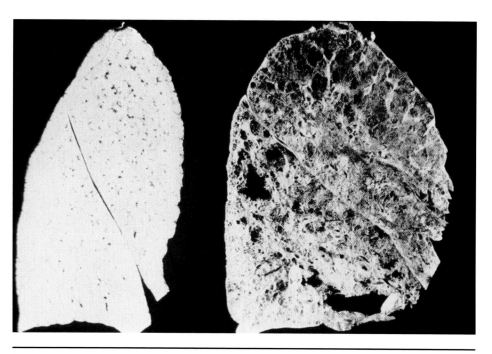

Lung sections of a nonsmoker (left) and smoker (right). © O. Auerbach/Visuals Unlimited.

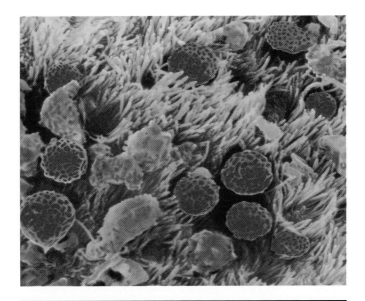

Trachea covered in allergens. Colored scanning electron micrograph of the surface of the trachea (windpipe) with breathed in pollen (orange) and dust (brown). These airborne particles may cause asthma or hay fever (allergic rhinitis). © Eddy Gray/Photo Researchers.

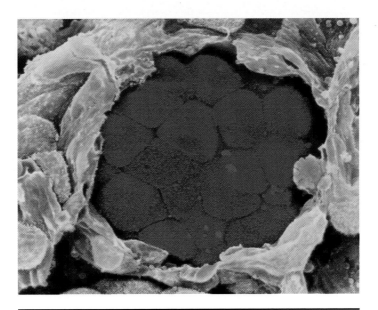

Lung cancer. Colored scanning electron micrograph of a tiny lung tumor (red) filling an alveolus (one of the blind-ended air sacs which make up the lungs). The individual cancer cells are coated with microscopic, hair-like structures known as microvilli. © Moredun Scientific/Photo Researchers.

Cilia in the lung bronchiole. © Dr. David Phillips/Visuals Unlimited.

5

Lung Disorders and Diseases

The lungs are unique among internal organs because they are continuously exposed to the external environment. This direct interface with the outside world results in the lungs being assaulted by numerous substances. As a consequence, the lungs are often the most likely organs to be affected by viruses and bacteria, allergy-causing pollen and dust, cigarette smoke, car fumes, and toxic chemicals from factories. Even a naturally occurring substance, like radon gas found in the soil and rocks, can harm the lungs.

Numerous diseases and conditions affect the lungs and impair their ability to provide the body with life-giving oxygen and rid it of carbon dioxide waste. Lung problems and diseases are usually classified according to three major categories, although many lung diseases, such as emphysema, often involve all three. They are:

- Obstructive lung diseases are those diseases in which the airways are narrowed or obstructed, thus decreasing the airflow.

- Restrictive lung diseases occur when the total volume of air that the lungs are able to hold decreases, usually as a result of a lost of elasticity in lung tissue or inability to expand the chest wall during inhalation.

- Diseases that affect the alveoli reduce their ability to diffuse oxygen into the blood.

According to the American Lung Association, lung diseases as a whole are the third most prevalent killer in America and are responsible for one in seven deaths. Each year, approximately 335,000 Americans die of lung disease. Although lung diseases affect all kinds of people, minority populations, especially African Americans, suffer from a disproportionate share of lung diseases, largely due to increased rates of cigarette smoking.

Ironically, despite the toll that lung diseases take on a person's health and the many deaths caused by them, a large majority of lung diseases could be prevented. For example, the number one cause of lung ailments is smoking cigarettes, a decision that is up to each individual.

Numerous diseases affect the lungs. Some of them are extremely rare. For instance, pulmonary nocardiosis is a lung infection caused by the fungus-like bacterium *Nocardia asteroides* and occurs in 500 to 1,000 people each year in the United States. This chapter discusses the following major or common lung diseases:

- Acute Respiratory Distress Syndrome (ARDS)
- Asthma
- Bronchitis
- Chronic obstructive pulmonary disease (COPD)
  - chronic bronchitis
  - emphysema
- Cystic Fibrosis
- Influenza ("the flu")
- Interstitial Lung Disease
- Lung Cancer
- Pleurisy
- Pneumonia
- Pulmonary Edema
- Pulmonary Hypertension
- Pulmonary Sarcoidosis
- Tuberculosis

ACUTE RESPIRATORY DISTRESS SYNDROME (ARDS)

Acute respiratory distress syndrome, or ARDS, is not a specific disease but rather a severe lung dysfunction that occurs rapidly and is associated with a variety of diseases or problems. These include pneumonia, shock, sepsis (severe body infection), and trauma, such as a severe blow to the chest, multiple fractures, and severe head injury. Also associated with ARDS are near drowning, smoke inhalation of toxic chemical and particulate matter, and overdoses of narcotics such as heroin or other drugs, including sedatives.

The most common physiological problems associated with ARDS are extensive lung inflammation and injury to the small blood vessels in the lungs. The prevailing symptom is severe shortness of breath that usually requires hospitalization. In nearly one-half of all cases, ARDS develops within twenty-four hours of the original illness or injury; the remaining 50 percent of ARDS cases almost all develop within three days.

Although many factors are associated with ARDS, scientists remain unsure as to exactly why it develops. To further cloud the clinical picture, most of the problems that can lead to ARDS are relatively common, yet not everyone who has any of the problems develops the syndrome. No matter how the lungs are injured, ARDS results when the lung's alveoli are damaged and eventually collapse, losing their ability to function in the gas exchange process. Other alveoli become filled by fluid, making the gas exchange process even more difficult. This interference can lead to respiratory failure. Fibrosis, or the formation of scar tissue, can occur in the lung, further interfering with the gas exchange process. Pneumonia sepsis and lung rupture can also occur, resulting in air leaking into surrounding areas.

Treatment: People who develop ARDS are hospitalized and treated with a mechanical ventilator and supplemental oxygen to help breathing while the lungs heal. Mechanical ventilation involves insertion of a tube, usually through the nose or mouth. A tracheotomy, in which an opening is cut through the neck and into the trachea and a tube inserted, is sometimes performed to ensure a safe airway or to avoid tracheal damage from a tube going down through the nose or mouth and into the trachea.

The lungs need six to twelve months or longer to completely recover from ARDS. The extent of recovery varies and often relies on how far ARDS has progressed and the problems or disease associated with its development.

ARDS patients may receive other therapies focusing on the illnesses or injury that has caused ARDS, such as antibiotics for bacterial pneumonia. Experimental therapies focus primarily on using a surfactant to replace the naturally occurring surfactant that keeps the alveoli open and functioning. The use of anti-inflammatory drugs is also being explored. Other drug therapies include medicines to maintain adequate blood pressure, pain relievers such as morphine, and anti-anxiety drugs to help the patient tolerate mechanical ventilation.

ASTHMA

According to some estimates, some 10 to 14 million Americans may suffer from asthma, with more than half of the cases occurring in children and teenagers. In fact, asthma is the most common chronic illness in childhood.

Asthma is sometimes referred to as "bronchial asthma" because it affects the bronchi, the small air tubes that branch off of the main bronchi and course throughout the lungs. The bronchi are surrounded by bronchial smooth muscle, which contracts or "twitches" as a defense mechanism in reaction to inhaled pollutants, irritants, and other factors. In the case of an asthma attack, these muscles essentially overreact to certain "triggers." The combination of muscle contraction, or spasms, along with bronchial inflammation, swelling, and excess mucus production make the bronchial airways so narrow that the individual finds it hard to breathe, especially to exhale air (see Figure 5.1). Asthma symptoms vary but usually include coughing or wheezing, shortness of breath, and a tightening in the chest.

Asthma is a serious condition and can be a medical emergency in cases of sudden, severe, and prolonged attacks. If the airways become totally blocked, respiratory failure, or suffocation, occurs because the body cannot get enough oxygen.

The triggers for asthma attacks vary from individual to individual. The most common causes are infections (primarily viral) and severe allergies to a wide variety of substances, from pollen and molds to house-dust mites

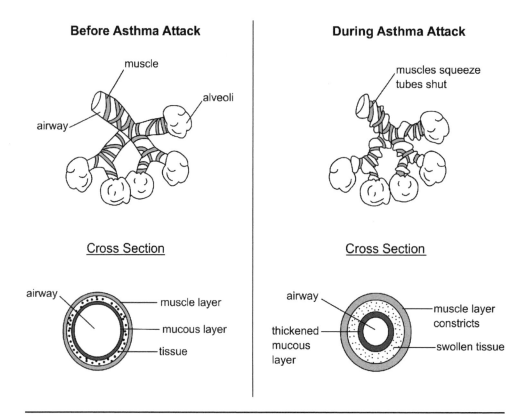

Figure 5.1. Before and during an asthma attack.

and certain foods. Various irritants, like tobacco smoke and chemical fumes and even cold air or exercise, can cause an asthma attack. In the case of exercise, rapid breathing through the mouth results in air bypassing the nose, which warms air entering the body. As a result, the air reaching the bronchial tubes is cold and triggers an attack in people who are overly sensitive to air temperature. Scientists have also shown that emotional factors like stress can trigger an attack or make it worse.

Treatment: Medical research has yet to find a cure for asthma, although some children who suffer from asthma seem to outgrow asthma attacks if not the disease itself. Nevertheless, many treatments are available for asthma. Anti-inflammatory medications help control swelling and inflammation while reducing sensitivity to triggers. Bronchodilator medicines relax the muscles surrounding the bronchi to open the air passages; these medicines are often the components of inhalers that asthmatics use when sudden attacks occur. Preventing an asthmatic attack is important. As a result, asthmatics should try to avoid the particular triggers that can cause their attacks.

BRONCHITIS

Bronchitis is an inflammation of the bronchial tubes. Most cases of bronchitis are acute, or short term, in nature and often follow an upper respiratory infection. Viruses, such as the influenza viruses and rhinoviruses, are the primary causes of these infections. In rare cases, a bacteria may be the culprit.

Whether bacterial or viral in nature, the infection leads to swelling of the bronchial tubes and excess mucus production, resulting in the characteristic cough associated with the disease (see Figure 5.2). In rare instances, breathing in environmental pollutants may cause acute bronchitis. Bronchitis that lasts for months is called chronic bronchitis and is primarily associated with other factors such as cigarette smoking. (Chronic bronchitis is discussed in detail in the next section.)

Although acute bronchitis is caused primarily by viral infections, smoking or other environmental pollutants can damage cilia in the bronchial tubes and increase the chances of developing bronchitis after an infection. The most common signs and symptoms of bronchitis include cough with mucus, shortness of breath, and wheezing. Acute bronchitis occurs in all age groups but most frequently in children under 5 years of age.

Treatment: Much like a viral upper respiratory infection, most cases of acute bronchitis resolve naturally after a few days to a week or so. Doctors recommend plenty of fluids and bed rest. In some cases, they may prescribe or recommend over-the-counter medications, such as cough suppressants, expectorants to relieve a dry cough, and a bronchodilator similar to those

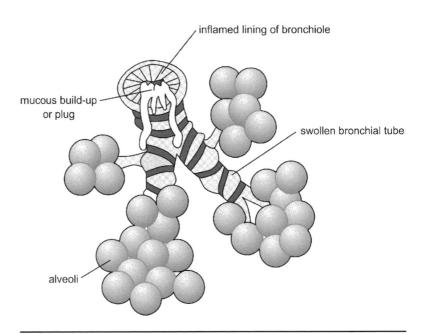

Figure 5.2. Inflamed bronchiole tube in bronchitis.

used for asthma. If a secondary bacterial infection has occurred, an antibiotic may be prescribed. If the symptoms of bronchitis persist over several weeks or months, the person may have a more serious problem, such as chronic bronchitis, asthma, or pneumonia.

CHRONIC OBSTRUCTIVE PULMONARY DISEASE (COPD)

According to some estimates, approximately 16 million Americans suffer from chronic obstructive pulmonary disease (COPD). The disease, which is characterized by reduced airflow through the respiratory system, is the fourth leading cause of death in the United States. By far, the primary cause of COPD is smoking tobacco. Other risk factors include exposure to dust and fumes (especially in the workplace such as mines), outdoor air pollution (associated primarily with people who smoke), repeated childhood respiratory tract infections, exposure to secondhand cigarette smoke, and some genetic deficiencies.

Although people with COPD often lead relatively normal lives for many years, COPD is a disease that progressively worsens over time. It is characterized by a chronic cough, spitting or coughing mucus, a loss of breath during exertion or exercise, and a growing inability to exhale air. COPD can encompass many conditions, including chronic asthma; the most common diseases associated with COPD are emphysema and chronic bronchitis.

Chronic Bronchitis

Unlike acute bronchitis, which originates mostly from viral infections, inflammation of the bronchial tubes in chronic bronchitis is primarily associated with cigarette smoking, which accounts for 80 to 90 percent of all chronic bronchitis cases. This increased incidence most likely occurs because inhaled tobacco smoke damages the cilia and impairs the body's ability to fight off infections.

Chronic bronchitis usually begins with inflammation in the smaller bronchial airways within the lungs, and then gradually spreads into the larger bronchial airways and tubes that make up the bronchial tree. The bronchial tubes have been irritated over a long period, and excessive mucus-producing cough occurs. The lining of the bronchial tubes thickens over time, which eventually can hinder airflow. Approximately 1,000 people in the United States die each year from chronic bronchitis.

Emphysema

Like chronic bronchitis, smoking is the overwhelmingly primary cause of emphysema. A deficiency of a protein known as alpha-I antitrypsin (AAT) can also lead to an inherited form of emphysema. Emphysema is the fourth leading cause of death in the United States and has risen by 40 percent since 1982.

Emphysema does not affect the bronchial tree but rather causes irreversible damage to the alveoli that cluster in sacs at the ends of the bronchial tree. Among the damages to alveoli are over-inflated alveoli, which can fuse with other alveoli to form enlarged alveoli (see Figure 5.3). As a result, the walls between alveoli are reduced in number and so are the blood vessels that course throughout these walls. This reduction of alveoli walls and surrounding blood vessels results in less surface area to provide for proper gas exchange and oxygenation of the blood. In addition, the surfactant that lines the alveoli within the lungs is damaged, leading to a loss of elasticity so that "stale" air left in the lung is never completely replaced by fresh air. The alveoli can eventually collapse, which causes air to become trapped and results in a greater difficulty in expelling air.

Unfortunately, most people do not pay attention to their symptoms of breathlessness until they lose 50 to 70 percent of their functional lung tissue. In addition to the common symptoms of COPD, other symptoms associated particularly with emphysema include weight loss and an increase in chest size called *barrel chest.*

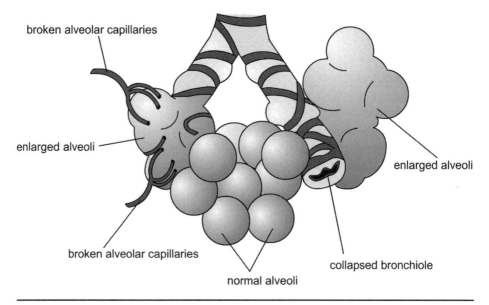

broken alveolar capillaries

enlarged alveoli

enlarged alveoli

broken alveolar capillaries

collapsed bronchiole

normal alveoli

Figure 5.3. Enlarged alveoli in emphysema.

Treatment: To date, medical science has yet to develop a treatment to cure or reverse the gradual worsening of COPD. The survival rate of COPD patients is not good. For example, six out of ten 60-year-old smokers with COPD will die before age 70. They are also four times more likely to die within ten years than a nonsmoking asthmatic.

The best way to prevent COPD is not to smoke. In most cases, COPD is not identified or diagnosed until patients reach their 50s and 60s. Most of the treatment focuses on treating the disease's symptoms and helping patients breathe easier and delay early death. The primary treatments prescribed for COPD include:

- corticosteroids (inhaled) to help reduce inflammation and mucus production
- surfactants to help replace the loss of natural surfactants in the bronchi and alveoli
- bronchodilators to help relieve airway constriction and reduce bronchospasms
- antibiotics to treat bacterial infections that often occur in COPD patients
- expectorants to loosen and expel mucus
- oxygen therapy for emphysema patients to help them oxygenate their tissues
- diuretics for patients who develop congestive heart failure

In some cases, lung reduction surgery may be performed to relieve symptoms and eliminate the need for as much oxygen. It involves making small incisions in the chest and inserting a tiny camera to view the lungs. Another incision is made through which the surgeon cuts portions of the lung allowing healthy lung tissue more room to expand. Lung transplantation is also performed in a small group of patients.

CYSTIC FIBROSIS

Cystic fibrosis (CF) is an inherited disease that affects the body's exocrine glands, which produce mucus, tears, sweat, saliva, and digestive juices. Caused by a defective gene, CF changes the chemical composition of these secretions, transforming them from thin and slippery to thick and sticky. Specifically, the gene changes a protein that regulates salt (sodium chloride) movement in and out of cells.

Although CF can affect the liver, pancreas, and the reproductive system, the disease mostly affects the respiratory and digestive systems. In the digestive system, the abnormal mucus can impede the digestive process. However, the disease's effect on the respiratory system is its most dangerous manifestation. The abnormal accumulation of thick mucus in the lungs sets up a breeding ground for bacteria, leading to many respiratory infections. It can also block the airways, and lung disease and respiratory failure are the usual cause of death. People with CF can also develop a collapsed lung, in which air leaks in to the pneumothorax (chest cavity).

Approximately 30,000 people in the United States have CF, which primarily occurs in white people of northern European ancestry. Between 2,500 to 3,200 babies are born with cystic fibrosis each year in the United States. The CF gene is a recessive gene, meaning that two copies of the gene must be inherited, one from each parent. As a result, someone who inherits only one copy of the gene will not develop CF or any symptoms of the disease. However, they do carry the gene and can possibly pass it on, but the disease will manifest itself only if the gene-carrier has a child with someone who also carries the gene. One in twenty-nine people in the United States is a carrier of the CF gene. If two "carriers" have a baby, the child has a 25 percent chance of getting CF and a 50 percent chance of being a carrier. There is also a 25 percent chance that the child will neither get the disease nor become a carrier.

The signs and symptoms of CF may vary, depending on what part of the body is most affected. In infants and young children, the most common signs and symptoms include foul-smelling and greasy stools or bowel movements, weight loss, breathlessness, wheezing, a persistent cough with a thick mucus, and numerous respiratory infections. In infants, older children, and adults, the most common symptom is salty sweat. As a result, a standard test for determining whether someone has CF is the sweat test, which measures the amount of sodium and chloride in a person's sweat. Most people with CF are diagnosed when they are infants or children. Other problems associated with CF include polyps (growths) in the nose, clubbing (enlargement and rounding) of the fingertips and toes, cirrhosis of the liver, and delayed growth.

Treatment: Although no cure or treatment exists for CF, many advances have been made over the years in treating the problems associated with the disease. As a result, the life expectancy of someone with CF has risen from five years in 1955 to a median survival age of 32 by the year 2000. (In statistics, the median is the value that separates the highest half of the sample from the lowest half.)

Treatments focus on managing the disease symptoms and include antibiotics for bacterial infections, decongestants and bronchodilators to open congested airways, mucus-thinning drugs, and nutritional regimens and supplements to provide proteins and calories to prevent malnourishment.

Since researchers discovered the specific defective gene that causes the disease, they have been investigating treatments involving gene therapy. This approach would involve delivering healthy, normal genetic material to cells in the airways. Another avenue of investigation is to modify the protein that the cystic fibrosis gene produces.

INFLUENZA ("THE FLU")

Influenza, more commonly known as "the flu," is a contagious disease caused by the influenza viruses. Much like the common cold, influenza spreads from person to person in several ways. The primary mode of infection is when a person is near an infected person who coughs and sneezes, or the uninfected person touches something that an infected person has coughed on or touched.

Unlike the common cold, however, the flu often causes severe and even life-threatening illnesses, including bacterial pneumonia. It also can exacerbate other medical conditions, such as asthma, diabetes, and congestive heart failure. Flu **epidemics** and **pandemics** have killed hundreds of thousands of people (see "The Flu of 1918"). Even people with the flu who do not suffer serious medical complications are much sicker than the common cold sufferer. In addition to the stuffy nose, sore throat, and dry cough that usually accompany a cold and the flu, people with the flu also suffer from fevers, headaches, body aches, and extreme tiredness.

Each year, approximately 10 to 20 percent of the people in the United States catch the flu. The influenza viruses also kill an average of 36,000 Americans each year and hospitalize another 114,000. Influenza, aided by its major complication of bacterial pneumonia, is the sixth most common cause of death in the United States.

The Flu of 1918

In 1918, a particularly virulent form of influenza traveled throughout the world, creating what is called a *pandemic*. The 1918 influenza pandemic, called the "Spanish flu," resulted in an estimated 20 to 50 million deaths. In the United States alone, the Spanish flu epidemic killed more than 500,000 people, including more U.S. soldiers than died in all of World War I. The flu was so pervasive that it reached all areas of the world, including remote northern frozen tundra where it wiped out entire Eskimo villages.

The Spanish flu had the ability to kill young and healthy adults, who in a matter of hours could develop fevers of 105°F and become so weak they could not walk. As the virus swept throughout the United States, fear became pervasive. Doctors were shocked when they performed autopsies on those who died and found a bloody and foamy liquid that entered the lungs in such quantities that it caused people to drown, or suffocate, in the mucus-like fluid. Another flu epidemic comparable to the Spanish flu has not occurred since 1918. If such an epidemic occurred today, an estimated 1.5 million Americans would die.

Since the pandemic of 1918, other pandemics have occurred. The "Asian flu" pandemic of 1957–1958 caused 70,000 deaths in the United States, and the "Hong Kong flu" killed approximately 34,000 Americans in 1968–1969. Potential pandemics that never developed include the "Swine flu" outbreak of 1976 and the "avian flu" outbreak of 1997, in which nineteen people in Hong Kong came down with a type of influenza infection that was thought only to occur in birds.

The potential for a new flu virus to emerge and result in a deadly epidemic that can quickly become a pandemic is very real, especially because of the growing number of people who travel word-wide in a matter of hours. Scientists are constantly researching influenza viruses in an effort to develop better vaccines. In the United States, the Centers for Disease Control and Prevention and local public health agencies also maintain surveillance systems that monitor such factors as weekly pneumonia and influenza deaths and overall influenza activity.

The influenza viruses are broken down into three major categories: type A, B, and C. The type A viruses are the most common and found in both people and many animals, including birds, pigs, ducks, and horses. Although people can catch the virus from animals, the virus is rarely spread this way. Type A viruses are the causes of the most serious epidemics and pandemics. The influenza B virus is also very contagious and causes large epidemics, but the disease and its complications are much milder than those caused by type A viruses. The influenza C viruses are much more mild than either A or B viruses and are not believed to cause epidemics.

Although influenza viruses infect all age groups and even people who are otherwise healthy, certain populations are at higher risk of catching the flu

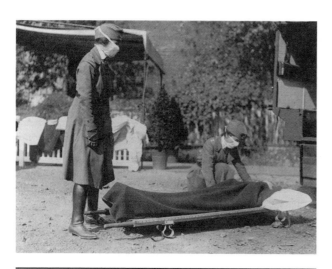

Nurses and a patient on a stretcher at the Red Cross Emergency Ambulance Station, Washington, D.C., during the influenza pandemic of 1918. Courtesy of the Library of Congress.

and developing serious complications. Those at the highest risk for serious complications are people who are 65 years of age and older, very young children, and anyone with a chronic medical condition like diseases of the heart, lungs, or kidneys.

Treatment: Because they are viruses, influenza cannot be treated by antibiotics, which work only against bacterial infections. For the most part, people with the flu recover in one to two weeks. Most people with the flu focus primarily on easing the symptoms by resting in bed, drinking lots of fluids, and taking over-the-counter medications such as decongestants for a stuffy, mucus-filled nose and pain relievers for body and muscle aches. Several antiviral medications are available but they do not cure the flu. For the most part, the antivirals reduce the infection's duration by a day or so. Some antivirals have been approved as a preventive step to avoid catching influenza.

By far, the best approach is to avoid the flu by getting a flu vaccination. Because flu viruses change over time, different vaccines are developed each year based on the viruses circulating at the time. Vaccines are made from killed, or deactivated, flu viruses grown in chicken eggs. However, even if a person receives a flu shot, they can still catch the flu because new strains of the virus may appear during the course of the flu season, which is when most flu cases occur because, it is believed, people are indoors more where they are in closer contact with other infected individuals and breathing recirculated air. Nevertheless, the flu vaccine usually leads to milder symptoms and complications even when it encounters a new viral strain. The flu vaccine can have side effects. Even though most of them are mild, certain people should consult their doctors and avoid vaccination, especially if they are allergic to chicken eggs, have previously had a serious reaction to a flu shot, have a paralytic disorder such as Guillain-Barré Syndrome, or are currently sick with a fever.

INTERSTITIAL LUNG DISEASE (PULMONARY FIBROSIS, INTERSTITIAL PULMONARY FIBROSIS)

Interstitial lung disease (ILD) encompasses more than 130 different lung disorders, the great majority of them chronic, or long-term, in nature. Although these various disorders differ in their cause and initial effects on the system, they all begin with an inflammation that can ultimately affect the connective tissue between the air sacs and blood vessels of the lungs. This tissue, called the *interstitium*, is where the gas exchange process between oxygen and carbon dioxide takes place.

The inflammation can start in three areas of the lung, and each type of inflammation is named accordingly. Bronchiolitis involves inflammation of the bronchioles, alveolitis of the alveoli, and vasculitis of the capillaries that surround the alveoli. In some instances, this inflammation may heal. In other cases, the inflammation may result in permanent scarring, or fibrosis, of the interstitial tissue, causing the lungs to lose elasticity. As a result, ILD is sometimes called *pulmonary fibrosis* or *interstitial pulmonary fibrosis*. If fibrosis occurs, the gas exchange process is permanently hindered, resulting in continued loss of the tissue's ability to transport oxygen. Depending on the extent of the fibrosis, this loss can result in various levels of disability in terms of breathing.

The most common early symptoms of ILD are difficulty breathing during exercise or exertion and an unproductive cough (no phlegm or mucus). As the disease progresses, other symptoms may include loss of appetite, fatigue, weakness, and weight loss. Among the numerous diseases that can cause ILD are sarcoidosis, **Wegener's granulomatosis,** and connective tissue or collagen diseases like rheumatoid arthritis and systemic sclerosis. Certain medications can cause pulmonary fibrosis, such as chemotherapy medications and, in rare cases, antibiotics. It can also be caused by radiation treatments for illnesses such as breast cancer. Exposure to environmental pollutants, including asbestos or metal dusts, can lead to fibrosis and ILD. In rare cases, ILD has been linked to families, meaning that in some instances it possibly may be an inherited genetic disease. When the source of ILD cannot be determined, it is called *idiopathic* ("of unknown origin") *pulmonary fibrosis*.

Treatment: Physicians consider many variables in treating ILD, including the cause and the severity of the symptoms, which can range from mild to extremely severe. Also of vital importance to treatment is early identification of the disease before lung damage becomes too great. For the most part, ILD treatments focus on the source of the problem if it is known and on easing the disease's symptoms and complications.

The primary medications used to treat ILD are oral corticosteroids, such as prednisone or methyl prednisone. Other medications, such as cyclophosphamide and azathioprine, are sometimes used if corticosteroid therapy is ineffective or the patient has serious side effects from the corticosteroids, such as high blood pressure. Patients whose breathing is severely impaired may require oxygen therapy, especially if they have low levels of oxygen in the blood. Influenza and pneumococcal pneumonia vaccines are often recommended to prevent infections and other respiratory problems.

Lung transplantation has also become a viable treatment for ILD and other lung diseases (see Chapter 6).

LUNG CANCER

Although lung cancer was a rare and virtually unknown disease in the United States prior to 1900, the incidence of lung cancer has been steadily rising in correlation with growing popularity of cigarette smoking since the 1930s. Not only is lung cancer one of the most common cancers in the United States, it is the leading cause of death due to cancer in American men and women. In 2002, about 150,000 people died from lung cancer, and approximately 170,000 new cases of lung cancer are diagnosed each year.

Lung cancer occurs when cells in the lungs begin to divide and grow abnormally. Once this process gets out of control, abnormal tissues called *tumors* are formed. Tumors can be either benign, meaning that they do not spread and are non-cancerous, or malignant, meaning that they are cancerous and will continue to spread, often throughout the body.

Although many types of lung cancer occur, they are grouped into two primary categories called non–small cell lung cancer and small cell lung cancer. This division is based on how the cancer looks under a microscope and not because of the tumor size. Non–small cell cancer is the most common form of lung cancer and includes squamous cell lung cancer, adenocarcinoma, and large cell carcinoma. These cancers usually spread more slowly than small cell lung cancer, which is also more likely to spread to other parts of the body.

Different types of lung cancer also affect different parts of the lung and respiratory system. For example, large cell lung cancer usually begins in the small bronchi of the bronchial tree, while squamous cell lung cancer most often originates in the large bronchi. Adenocarcinoma often begins at the outer edges of the lung. Small cell lung cancer is usually found in the bronchial mucosa, a layer of tissue beneath the epithelium. The primary symptoms of lung cancer are a continuing cough that worsens over time, repeated chest infections, chest pain, coughing up blood or blood stained phlegm, breathlessness, loss of appetite and/or weight, tiredness, and swelling of the neck and face. Because many of these symptoms are associated with other medical problems, only a doctor can correctly diagnose lung cancer.

Smoking cigarettes causes most cases of lung cancer (see "Did You Know? Cigarette Smoking"). Cigar smoking also causes lung cancer, but not as often because people usually smoke less cigars and do not inhale cigar smoke as deeply. More than 90 percent of all lung cancer deaths are related to smoking tobacco and inhaling the more than 4,000 chemicals that tobacco contains, many of which are **carcinogens**. In addition, many lung cancer deaths are due to people being exposed to secondhand environmental smoke, that

Did You Know? Cigarette Smoking

Cigarette smoking is the single most preventable cause of premature death in the United States and accounts for one out of every five deaths, killing more than 430,000 people in the United States each year. Compared to people who do not smoke, men who smoke increase their risk of dying from lung cancer by more than twenty-two times and from chronic bronchitis and emphysema by nearly ten times. Women, in whom lung cancer increased dramatically between 1960 and 1990, have a twelve-times-increased risk of dying from lung cancer and a ten-times-increased risk of dying from chronic bronchitis and emphysema. Exposure to secondhand, or environmental, tobacco smoke also causes an estimated 3,000 deaths from lung cancer in American adults who do not smoke. Maternal smoking has also been strongly associated with adverse respiratory effects in children, and there is evidence that it even affects the child while it is in the womb (see Chapter 8 for more on smoking).

is the smoke blown into the air by smokers. Not everyone who smokes gets lung cancer, and some people develop the disease because of other reasons, including exposure to radon, asbestos, and certain air pollutants and other substances. Lung diseases like tuberculosis also increase the risk for developing lung cancer.

Treatment: The primary treatments for lung cancer are surgery, **chemotherapy,** and radiation therapy. Usually a treatment's effectiveness depends on how early the cancer has been diagnosed and how far it has advanced.

The types of lung cancer surgery are a resection, or removal of only a small part of the lung; a lobectomy, in which one the lobes of the lungs is completely removed; and a pneumonectomy, or the removal of the entire lung. In cases of non–small cell lung cancer, cryosurgery may be used in which the cancerous tissues are frozen and destroyed.

Chemotherapy is the use of anticancer drugs to disrupt cancer cell growth and may be used instead of, or in addition to, surgery. Most chemotherapeutic agents are injected intravenously but some are in pill form. Radiation therapy, sometimes called *radiotherapy*, uses high-energy rays focused on a small area to kill only cancer cells. It can be used before surgery to shrink a tumor and make it easier to remove, and after surgery to destroy any cancer cells that may remain in the body.

PLEURISY

Pleurisy, sometimes referred to as *pleuritis*, is an inflammation of the pleura, the transparent, moist, double-layered membrane that lines the lung.

Usually a sign of another illness, pleurisy has many causes, including bacterial and viral infections of the lungs, tuberculosis, blood clots, lung cancer, and lupus. Other causes include injury or trauma, such as a rib fracture, and air pollutants, such as asbestos.

The primary symptom of pleurisy is a stabbing chest pain that is aggravated by breathing and coughing. Sometimes a person with pleurisy will also feel pain in the shoulder, neck, or abdomen. Other symptoms, which depend on the severeness of the pleurisy and its underlying causes, include difficulty breathing, dry coughing, headache, chills, and fever.

The two primary types of pleurisy are called *wet pleurisy* and *dry pleurisy*. In the more common dry pleurisy, the inflamed pleura, which is normally smooth and protects and lubricates the lungs as it inflates and deflates, develops fibrous adhesions between the two membrane surfaces. As a result, the pleura becomes rough so that breathing and other chest movements result in the surfaces rubbing together and creating pain. In most instances, the person feels the pain in a very exact location in the chest. This rubbing also produces a "grating" sound that can be heard with a stethoscope and sometimes by putting an ear to the person's chest.

Wet pleurisy can cause fluid buildup in between the double-layered pleural membranes, causing a condition known as *pleural effusion*. The two pleural membranes normally have only a thin lubricating layer of fluid between them so that they fit tightly together, like a balloon with no air in it. Although pleural effusion can ease the pain of pleurisy, it also restricts the lung movement and can make it more difficult to breathe.

Treatment: Often, pleurisy will go away on its own, but it largely depends on the underlying cause or causes. For example, if the cause is a bacterial infection, a doctor may prescribe antibiotics. Other medications often prescribed include anti-inflammatory medicines, cough syrup with codeine, and/or pain medications like aspirin.

If excessive fluid has built up in the pleura, the individual may have to be hospitalized and have the fluid withdrawn through a needle or tube inserted into the chest. The chest pains associated with pleurisy can be controlled to a certain extent by limiting movement, including lying in a special way so that the breathing movement is limited to reduce stretching of the sore tissues.

PNEUMONIA

Pneumonia is an inflammation of the lungs caused by an infection or an injury, such as the inspiration of dust or chemicals. Specifically, the alveoli fill with fluid, such as mucus or pus. As a result, oxygen transport throughout the body via the blood is impaired and cell damage can occur. Approximately two to three million people each year develop infectious pneumonia in the

United States, resulting in 40,000 to 70,000 deaths due to restricted oxygen supply, cell damage, and infection spreading to other organs in the body.

More than thirty kinds of pneumonia are known, ranging from mild to life threatening (see "SARS: A Deadly Respiratory Disease Emerges in the Twenty-First Century"). Pneumonia that affects entire lobes of the lungs is

SARS: A Deadly Respiratory Disease Emerges in the Twenty-First Century

On February 11, 2003, the World Health Organization (WHO) received the first official notice of an outbreak of atypical pneumonia in China's Guangdong Province. Approximately 30 percent of the cases were reported in healthcare workers. Ten days later, an infected doctor who had treated patients in China traveled to Hong Kong and stayed at a four-star hotel. Within days some hotel guests and visitors began to come down with the "mystery" disease, and similar cases soon turned up in nearby Vietnam and Singapore. Almost simultaneously, the disease began to spread to various cities around the world, including Toronto, Canada.

Dr. Carlos Urbani, a WHO epidemiologist in Hanoi, Vietnam, first identified the unknown respiratory disease on February 28. The disease was named severe acute respiratory syndrome, or SARS, and its symptoms were similar to the flu and pneumonia and included a high fever, headache, an overall feeling of discomfort, and body aches. After two to seven days, dry cough and trouble breathing occurred. As an infectious disease, SARS spread through contact much the same as other infectious respiratory diseases. On March 29, the 46-year-old Dr. Urbani died of SARS in Thailand.

The health care community quickly recognized SARS as a particularly serious threat because they had no known treatment or vaccine. To stop the spread of SARS, public health organizations resorted to age-old techniques of isolating patients and quarantining populations that may have been exposed to others with SARS. Eventually, scientists identified the virus that causes SARS as a previously unrecognized coronavirus, which is a family of viruses notorious for frequent mutations.

On July 5, 2003, WHO announced that the human chain of SARS transmission appeared to be broken and that the SARS outbreak had been contained worldwide. Nevertheless, the SARS virus had traveled in humans to thirty countries, becoming deeply embedded in six. As of July 2003, SARS had infected 8,439 people, and 812 people died from the disease.

Despite the headway made against SARS, WHO warned that new cases may appear and that much has to be learned about the disease. For example, based on knowledge about other coronaviruses, SARS may be seasonal and return. Furthermore, the original source of SARS is unknown and may still be in the environment, harbored, for example, by a group of animals.

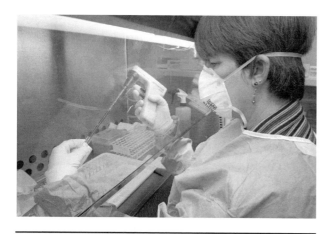

Processing SARS specimens in the laboratory, 2003. © Centers for Disease Control and Prevention.

called *lobar pneumonia*. Bronchial pneumonia affects patches throughout the lungs (see Figure 5.4). Although pneumonia can have many causes, the primary causes are bacteria, viruses, mycoplasmas (the smallest known living agents of disease, which have characteristics of both viruses and bacteria), fungi, and various chemicals. *Pneumocystis carinii* pneumonia, believed to be caused by a fungus, is particularly virulent in people with weakened immune systems, such as AIDS patients.

Other rare pneumonias are caused by inhaling certain food, liquids, gases, dust, or other foreign bodies. Bronchial obstructions, such as a tumor, can also cause pneumonia.

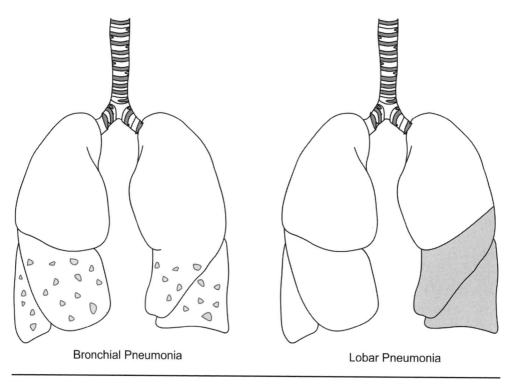

Bronchial Pneumonia   Lobar Pneumonia

Figure 5.4. Bronchial and lobar pneumonia.

Rickettsia, which is another organism considered to be something between a virus and a bacterium, causes many diseases that can affect the lungs, like Rocky Mountain spotted fever, Q fever, typhus, and psittacosis. An extremely dangerous type of pneumonia if not treated early is tuberculous pneumonia.

The symptoms of pneumonia may vary according to the type of pneumonia. For example, mycoplasmal pneumonia is notorious for causing a sore throat and headache in older children and adolescents. Bacterial pneumonia tends to cause illness more quickly with a sudden onset of high fever and rapid breathing, while viral pneumonia is more gradual in its progression and often produces less severe symptoms. The most common signs and symptoms of pneumonia are a fever, chills, cough, rapid breathing, labored breathing, vomiting, chest and/or abdominal pain, loss of appetite, fatigue, and a bluish cast to lips and fingernails. Some or all of these symptoms may be present depending on the type and progression of the pneumonia.

Pneumonia is also classified according to where people are exposed to the disease. For example, *community-acquired pneumonia* means that the pneumonia is caught during normal daily activities in the community, such as going to work or school. *Hospital-acquired pneumonia,* also called *nosocomial pneumonia,* primarily affects hospital patients who are in intensive care units or have compromised immune systems. *Aspiration pneumonia* results from inhaled particles reaching the lungs, usually as a result of stomach contents reaching the lungs after vomiting. Finally, pneumonia that attacks people with compromised immune systems is called *opportunistic pneumonia.*

Although anyone can catch pneumonia, there are certain factors that increase a person's risk of getting sick and developing serious complications. People 65 years old and older are at greater risk for both contracting and dying from pneumonia, primarily because their immune systems are generally weaker (see "Did You Know? William Henry Harrison"). In a twenty-

Did You Know?
William Henry Harrison

Pneumonia killed William Henry Harrison, the ninth president of the United States. Harrison came down with pneumonia during his inauguration in 1841. He died only thirty-one days after he took office.

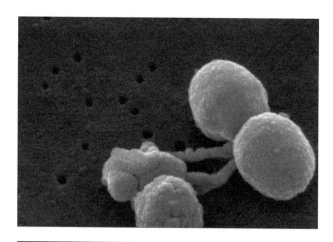

A photomicrograph of *Streptococcus pneumoniae.* © Centers for Disease Control and Prevention.

year study, patients over 80 years old had the highest mortality rate from one of the most virulent forms of pneumonia, called *pneumococcal pneumonia,* which is caused by the *Streptococcus pneumoniae* bacteria (see photo). Pneumonia also occurs often in very young children because their immune systems have not yet fully developed. People with immune deficiency diseases such as AIDS or with chronic illnesses like cardiovascular disease or emphysema are also at high risk. Other risk factors include removal of the spleen, long-term use of **immunosuppressive drugs,** smoking tobacco, drug and alcohol abuse, and agriculture and construction workers exposed to chemical fumes and other airborne particles.

Treatment: Antibiotics are the treatment of choice for people with bacterial pneumonia and mycoplasmal pneumonia. However, the number of new bacterial strains with resistance to current antibiotics is rising. Only six percent of the bacteria that caused pneumonia were resistant to antibiotics in 1994. Over the next six years, that number had risen to thirty-four percent. Antibiotics are not effective against viral pneumonia. Although some anti-viral medications have been developed, they are not very effective against pneumonia.

For most cases of viral pneumonia, and as a supplement to antibiotic treatment for bacterial pneumonia, physicians recommend plenty of rest and drinking lots of fluids. They may also recommend over-the-counter medications such as pain relievers and anti-inflammatory medications. Cough suppressants may be prescribed in some cases but usually are avoided because coughing helps to clear the lungs. Serious cases of pneumonia may require hospitalization.

Pneumonia can be prevented. Because many of the sources of pneumonia are contagious infections, standard hygienic practices can help a person avoid pneumonia, including avoiding people who have a cold or flu, washing hands often to eliminate germs that may have landed on doorknobs and other objects from people who have infections, and not smoking tobacco. Since pneumonia is often a complication of influenza, or the flu, getting vaccinated against the flu is one of the primary ways to prevent catching pneumonia.

PULMONARY EDEMA

Pulmonary edema is a good example of how the various body systems are interrelated and how problems in systems can affect each other. Although pulmonary edema can have various causes, it usually results from malfunctions in the circulatory system, such as congestive heart failure and cardiomyopathy, or stiffening of the heart muscle. As a result, a person's heart loses the ability to properly pump blood throughout the body, which increases pressure in the pulmonary veins that run throughout the lungs. As this pressure increases, fluid from the pulmonary capillaries leaks, or is pushed, into the alveoli. This leaked fluid hinders the gas exchange process, resulting in shortness of breath. If left untreated, a person can become starved for oxygen and literally suffocate. Pulmonary edema also has other causes, including severe infections, excess body fluid caused by kidney failure, and direct injury to the lungs from toxins.

High-altitude pulmonary edema (HAPE) occurs most often when people travel to high altitudes, normally 8,000 feet and up, although it can occur at lower altitudes. The reasons for HAPE are not fully understood, although it is exacerbated by exercise. One hypothesis has been that higher altitudes cause swelling in the pulmonary vessels within the lungs. Another possible cause is that the capillaries became leaky due to inflammation. Recent studies indicate that HAPE may be due to the lack of oxygen at higher altitudes causing lower pulmonary capillary pressure. HAPE is the most common cause of death related to altitude sicknesses.

The most common symptoms of pulmonary edema are shortness of breath and difficulty breathing, a feeling of "air hunger" or "drowning," grunting or gurgling sounds with breathing, and wheezing and a cough. Pulmonary edema may also induce anxiety, restlessness, excessive sweating, and pale skin. People may also cough up blood and become confused. A rapid heartbeat often occurs.

Treatment: The primary treatment for pulmonary edema is supplemental oxygen via a face mask or nasal prongs. A breathing tube connected to a ventilator may have to be placed in the windpipe (intubation). Often diuretics and other drugs are given intravenously to remove fluid via the urine and relieve congestion.

In most cases of pulmonary edema, treating the heart condition is essential. Doctors often prescribe medications designed to strengthen the heart and relieve heart pressure. In the case of people with HAPE, moving the person to a lower altitude is essential. If they are not able to receive emergency treatment with oxygen, vasodilator medications are often used.

PULMONARY HYPERTENSION

Although the true incidence of this extremely rare lung disease is un-known, researchers estimate that pulmonary hypertension occurs only once or twice per one million people. In the United States, doctors diagnose ap-proximately 300 new cases of pulmonary hypertension per year.

Because of its rarity, relatively little is known about the cause of pul-monary hypertension. However, the disease's effects are well documented. Pulmonary hypertension results in extremely high blood pressure in the pulmonary artery, which is the blood vessel that carries oxygen from the heart's right ventricle to the miniscule blood vessels, or pulmonary capil-laries, that course throughout the lungs. Ultimately, the pulmonary arteries thicken, causing restricted blood flow through the lungs and sometimes cre-ating blood clots. As a result, the right ventricle begins to work harder to make up for this restricted flow. The extra strain can result in a heart attack and death. The primary symptoms of pulmonary hypertension include shortness of breath, excessive fatigue, dizziness, fainting spells, weakness, swelling in the ankles or legs (edema), and chest pain (angina).

Although science has yet to determine the cause or causes of pulmonary hypertension, researchers have determined that different types of pul-monary hypertension are associated with various problems and diseases. *Pulmonary arterial hypertension* features changes in the pulmonary arter-ies of the lung and has been associated with collagen vascular disease, HIV infection and AIDS, and various drugs or toxins, including diet pills, am-phetamines, and cocaine.

Another type of pulmonary hypertension, called *pulmonary venous hyper-tension*, is characterized by difficulty in blood flowing out of the lungs. It has been linked with problems in the left side of the heart and tumors or cancer in the chest wall. Known disorders of the respiratory system, such as inter-stitial lung disease and chronic obstructive pulmonary disease, have also been associated with pulmonary hypertension. The disease has been found to occur in some families, indicating that it may have genetic causes in some cases.

Treatment: Prior to 1990, when few treatments were available for pul-monary hypertension, the survival rate for patients was two to four years. However, numerous treatments based on the type of pulmonary hyperten-sion have become available primarily to treat symptoms and slow the dis-ease's progression. In rare cases, when the disease is caught early, damage done to the lungs and heart can be reversed. Drug therapies include anti-coagulants to prevent blood from clotting, diuretics to help control blood pressure and stop fluid accumulation in the legs and ankles, and vasodila-tors to open blood vessels. Many pulmonary hypertension patients also re-ceive oxygen therapies to help them breathe and help their bodies to acquire enough oxygen.

Various surgical treatments are also available for pulmonary hypertension. A thromboendarterectomy can be performed to clear blood clots that form in the pulmonary blood vessels over an extended period of time. A rarely performed procedure for only the most seriously ill patients is an atrial septostomy, which involves cutting a hole in the heart between the right and left atria to reduce the pressure on the right side of the heart. Another surgical procedure is a double heart-lung transplant, which is primarily used as a last resort in patients who are extremely ill.

PULMONARY SARCOIDOSIS

Although it can occur in almost any organ, ninety percent of pulmonary sarcoidosis cases occur in the lungs. Sarcoidosis, which stems from the Greek words *sark* and *oid*, together meaning "flesh-like," results from a type of inflammation that progresses to form small lumps, or granulomas, on the affected tissue. Once thought to be an extremely rare disease that occurred primarily in Scandinavian countries, physicians in the United States identified a large number of cases in the mid-1940s during the course of chest x-ray screening for the military. Because sarcoidosis often goes undetected, the best estimates for its occurrence in the United States is 5 in 100,000 white people and 40 out of 100,000 black people.

Sarcoidosis of the lungs is believed to begin with inflammation of the alveoli (alveolitis). If the inflammation does not clear up spontaneously, granulomas form and can eventually lead to fibrosis, causing the lungs to stiffen. Although the cause is unknown, scientists believe sarcoidosis of the lung and other types may be due to an irregular immune response to some type of trigger, such as a chemical, drug, or virus. If damage to lung or other organs is significant, sarcoidosis can be fatal.

Sarcoidosis can occur at any age but usually appears between the ages of 20 and 24. Symptoms vary from person to person; the most common symptoms are shortness of breath, a persistent cough, skin rashes (on face, arms, or shins), eye inflammation, weight loss and fatigue, night sweats, and fever. Because people with sarcoidosis often do not show any signs or symptoms, many cases probably go undetected.

Treatment: Approximately 60 to 70 percent of sarcoidosis cases go away spontaneously within two years of symptoms developing. However, 10 to 15 percent of the patients suffer from chronic sarcoidosis, and approximately 20 to 30 percent of these cases result in permanent lung damage. Treatments focus on keeping the lungs and other affected organs operating properly. The primary drug therapy is corticosteroids. Symptoms often reoccur, so patients may have to continue corticosteroid treatment for many years to prevent relapse. Immunosuppressives and other miscellaneous drugs have been tested and used, but none have proven as effective as corticosteroids. No treatment has been found to reverse fibrosis due to advanced sarcoidosis.

TUBERCULOSIS

A bacterial infection of the lungs and occasionally other parts of the body, tuberculosis (TB) is caused by the *Mycobacterium tuberculosis*. This bacterium has been killing people for thousands of years as evidenced by its discovery in Egyptian mummies over 4,000 years old. After the first effective antibiotics were developed for TB in the 1940s and 1950s, the medical community began to gain control of the disease. By the 1980s many believed that the disease would one day be vanquished from the face of the earth. However, TB has made a resurgence with eight million people developing the disease each year throughout the world and three million dying (see "What Caused TB Resurgence in the United States?").

Tuberculosis is an airborne infection passed on much like the cold and flu, primarily from person to person via tiny droplets forced into the air when a person sneezes, coughs, talks, or even laughs. A person can be infected with the bacterium but never develop the disease, especially if they are healthy and have strong immune systems. Only people who actually become sick are contagious—that is, they have an active, not a latent, form of the disease. The World Health Organization estimates that only about 5 to 10 percent of healthy people exposed to tuberculosis bacteria will actually develop active TB, sometimes months or years after initially contracting the bacteria if their immune systems become damaged. However, sick people whose immune systems are weak or compromised are much more likely to develop active TB quickly.

When a person develops active TB, they begin to show symptoms such as loss of appetite and weight loss, fever,

A poster from the 1920s warns about the spreading of tuberculosis and influenza. © National Library of Medicine.

What Caused TB Resurgence in the United States?

After the discovery of effective antibiotics against TB, the disease in the United States began a steady decline that lasted until about 1985. At that point TB cases began to rise again. Numerous factors have been associated with its resurgence. They include:

- Appearance of the HIV/AIDS epidemic, which increased susceptibility in people who contracted HIV/AIDS

- Increase in poverty and homelessness, because TB transmission occurs easily in crowded shelters where people are weak, due to poor nutrition, alcoholism, and drug addiction, and where people are often exposed to others with TB

- Increase in injection drug use, because people are likely to share needles with someone who has the TB bacterium

- Increase in people living in the United States who are from foreign countries where TB occurs often, including African, Asian, and Latin American nations

- Failure of patients to follow their prescribed antibiotic regimen, leading to reoccurrence of TB and multidrug resistant TB, or MDR-TB

- Increase in long-term care facilities where many of the residents, who may have had a latent form of TB since their youth and then develop an active form due to poorer health, spread the disease to other residents and health care workers

Fortunately, the number of tuberculosis cases in the United States has been on the decline since 1994, largely due to demographic changes and increased TB control efforts, in particular ensuring that people with TB follow through on their drug regimens. Nevertheless, TB rates have continued to rise outside of the United States, largely in poorer undeveloped regions where treatment and funding for treatment are not always easy to obtain, especially for MDR-TB.

night sweats, and coughing (including coughing up blood). If TB spreads unhindered, it can cause death. Tuberculosis causes death not by suppressing the immune system but by activating an overwhelming **inflammatory immune response** that actually damages and destroys the tissue.

Regardless of race or nationality, age, or socioeconomic status, anyone can get TB when they are around someone who already has it. However, some people are at higher risk than others. In addition to people who are already sick and/or have weakened immune systems due to disease such as AIDS,

Inhalation Anthrax and Terrorism

The acute infectious disease known as anthrax is extremely rare in humans. However, growing concerns over bioterrorism, such as the intentional spread of anthrax through letters in the United States in 2001, have aroused public awareness of the disease. Anthrax is caused by the spore-forming bacterium *Bacillus anthracis* and is most commonly found in hooved animals, such as cows, sheep, camels, and antelopes. Anthrax can infect humans when they are exposed to infected animals, infected animal flesh, or breathe air laden with spores of the bacilli. It usually occurs in people whose occupations expose them to anthrax, such as veterinarians and wool sorters.

There are three types of anthrax, distinguished by their modes of transmission. *Cutaneous anthrax* occurs when the bacteria enters the skin, for example, via a cut or abrasion. *Gastrointestinal anthrax* in humans occurs primarily from consumption of contaminated meat. The spread of anthrax through meat products has been largely prevented since the first successful immunization of livestock occurred in 1880. The third type of anthrax infection has raised the most concern related to bioterrorism. Called *inhalation anthrax*, it is contracted from inhaling thousands of tiny bacteria spores.

The initial symptoms of inhalation anthrax are similar to the common cold and flu, including a sore throat, mild fever (greater than 100°F), shortness of breath, cough, and muscle aches. If left untreated it can progress to severe breathing problems and shock. Inhalation anthrax is often fatal. According to the Centers for Disease Control and Prevention (CDC), the fatality rate for inhalation anthrax could be as high as 75 percent, even when supportive care is provided. (The CDC notes that this statistic is based on incomplete information.) Fortunately, unlike the cold or flu, anthrax is extremely unlikely to spread from person to person.

Anthrax can be detected through several methods, the most common is a nasal swab test that takes a sample from inside the nostrils. The disease can be treated successfully with antibiotics if they are taken during the disease's initial stages. Several antibiotics are effective against anthrax, including ciprofloxacin, doxycycline, and amoxicillin given over a sixty-day period. Antibiotics can also be used as a preventive measure but are not recommended unless people are exposed to anthrax.

According to the American Medical Association (AMA), prescribing unnecessary antibiotics for a potential threat of anthrax contamination in the general public can have serious repercussions. Of most concern is the development of or-

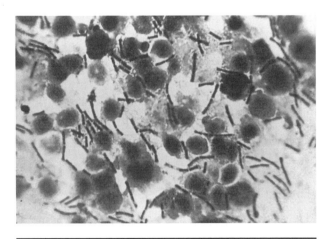

Bacillus anthracis—anthrax spores. © Centers for Disease Control and Prevention.

ganisms that are resistant to antibiotic treatment, which could lead to more disease and death in the public from respiratory and other infections. Furthermore, if a bioterrorism event occurred involving anthrax, antibiotics for preventing the disease are stockpiled in several areas in the United States and can be delivered to any airport within twelve hours. A vaccine is also available to prevent infection, but it is recommended only for those who have a high risk of being exposed to the disease.

---

diabetes mellitus, and cancer of the head or neck, people who work in long-term and residential care facilities such as nursing homes, prisons, and some hospitals are at higher risk.

Malnourishment also increases a person's risk of contracting TB. The risk increases among people who belong to certain groups in which the prevalence of TB infection is high: people who were born in areas with a high prevalence of TB; medically underserved populations, including the poor and the homeless; certain racial and ethnic minority groups; and intravenous drug users. If a person is thought to have TB, a TB skin test is available.

Treatment: Tuberculosis can be cured in nine out of ten patients with appropriate antibiotic treatment, usually involving several different antibiotic drugs given over a period of six to twelve months. Unfortunately, many patients do not complete their drug regimen once they start to feel better. As a result, they can become sick with TB a second time, and the TB may be harder to treat because it has mutated or changed upon exposure to the first round of antibiotics. As a result, it has become drug resistant. Known as *multidrug resistant tuberculosis*, or MDR-TB, these highly mutated forms of TB require special drugs, many of which can cause serious side effects. Furthermore the drug regimen may have to be continued for up to two years. Even with such treatments, four to six out of ten patients with MDR-TB die.

Tuberculosis is preventable. For example, the drug isoniazid can be given to people who have the latent form of TB to prevent it from developing into active TB. A vaccine is also available for TB called the *Bacille Calmette-Guerin* (BCG) vaccine. It has been widely used in countries with a high incidence of TB. Unfortunately, the more than 80-years-old vaccine has been found to be increasingly ineffective, especially for adults with TB infection of the lung. In addition, the vaccine interferes with the TB skin test, so that it becomes difficult to determine if someone has been infected with the TB bacteria.

Scientists are trying to develop new vaccines for TB. In the case of hospitals, clinics, and long-term care facilities where TB can easily spread, preventive measures include using ultraviolet light to sterilize the air and special air filters, respirators, and masks. Patients with TB are often isolated in rooms with controlled airflow and ventilation.

Monitoring TB patients and ensuring that they complete their drug regimen is also too expensive and difficult in some countries. As a result, TB kills more young people and adults around the world than any other infectious disease. The World Health Organization predicts that by 2020 as many as one billion people will be newly infected with TB, resulting in 70 million deaths.

Treatments: Past, Present, and Future

Although early efforts to treat respiratory illnesses and cure respiratory diseases were often unsuccessful, occasionally remedies worked and had a sound scientific basis. For example, records indicate that effective herbal remedies for asthma date back five millennia to the Chinese emperor Shennung. Sumerian physicians knew how to use opium to help mask pain, and ancient Egyptians used herbs to help relieve the symptoms of respiratory and other ailments. The ancient Greeks had also established a substantial *pharmacopoeia*, or collection of drugs, based on plants and plant derivatives. The Romans were known for their use of purgatives to empty the bowels.

The basis for vaccines was also established many centuries ago in China when dried crusts of smallpox scabs were ingested or applied to the skin and nasal mucous membranes to fight off the then-prevalent disease. Ancient doctor-philosophers, such as Hippocrates (460–377 BCE), also were aware of the basic principles of infectious diseases and epidemics. By Biblical times, governments were using quarantines to isolate people who were considered infectious.

Over the centuries, many remedies became folklore or "old-wives tales." Colds and coughs were fought using poultices (soft, usually heated, and sometimes medicated substances spread on a cloth) and teas made from herbs, and wild cherry and honey were used to reduce phlegm and coughing. Although some "folk" remedies were simply desperate measures without any basis to support their effectiveness, many others resulted from a trial and error approach. As a result, modern science has discovered that some of these remedies are effective because of natural **antihistamines** and other substances found in them.

Most early treatments, however, were based on rudimentary knowledge about plants and other substances, heavily spiced with a mixture of religious and philosophical thought. Aromatherapy, for example, dates back at least to 4500 BCE, when various substances with aromatic properties were used for both religious rituals and medicinal applications. Nevertheless, agents such as menthol, eucalyptus, frankincense, and balsams have long held a favored status as respiratory therapies that soothe inflamed mucosa, alleviate nagging coughs, and enhance expectoration. Once again, science has shown that these treatments can have some positive effects on treating the symptoms of respiratory problems, as is evidenced by the use of mentholated cough drops.

Modern medicine, of course, has come a long way in providing treatments based on hard scientific evidence and years of rigorous research. But the majority of these advances didn't take place until the twentieth century. Prior to that time, most physicians and scientists were too busy just trying to figure out the causes of the most common diseases, including infections and colds.

A turning point in successfully treating many diseases occurred in 1857, when Louis Pasteur (1822–1895) realized that some microorganisms could kill other microorganisms; he later applied active immunization by using attenuated (weakened) anthrax cultures to promote a reaction against anthrax. By the late nineteenth century, the germ theory for many diseases had also been established. As a result, medical science began moving into the realm of accurate diagnosis and cure, as well as prevention.

According to many, however, the golden age of medicine didn't really begin until the 1930s when a German pharmacologist, Gerhard Domagk (1895–1964), discovered that a dye used to color cloth appeared to cure streptococcal infections in mice. He then used the dye as an injection to cure his daughter who was dying from a strep infection. The discovery quickly led to treatments for pneumonia and other bacterial infections.

This chapter focuses on some of the major advances made in treating respiratory diseases over the years, such as the discovery and use of antibiotics. It will also look at some current state-of-the-art therapies, such as lung transplantation, Finally, the chapter discusses future treatments for respiratory diseases, like aerosolized drug delivery to the lungs and gene therapy.

OXYGEN THERAPY

After the discovery of oxygen in the 1740s and the realization that it most likely played a vital role in respiration and body function, doctors and scientists soon began pondering the use of oxygen as a form of therapy. Oxygen was first used in France in 1783 as a possible treatment for tuberculosis and **asphyxia** in infants. In the late 1790s, Thomas Beddoes (1760–1808) founded the Pneumatic Institution for Inhalation Gas Therapy and took part in what were early versions of clinical trials to see if oxygen was therapeutic for diseases, including tuberculosis (then known as consumption), asthma, and even venereal disease.

Although many claims were made for the value of oxygen therapy over the years, little concrete scientific data was gathered to prove its effectiveness for any disease, including respiratory diseases. As a result, most physicians believed it to be ineffective. Nevertheless, home oxygen delivery methods were available to patients in New York City as far back as the 1870s.

One of the primary problems of early oxygen therapies was the common approach of not supplying oxygen continuously but on an intermittent and haphazard basis. For example, an intermittent positive pressure breathing device in 1917 released oxygen twelve times per minute for the patient to breathe in, and the therapy was provided only a few times a day. Furthermore, the equipment was often faulty, and masks were often held away from the face; they also tended to leak. The medical community had little or no understanding at the time of the need for continuous oxygen supply.

The modern era of oxygen therapy began after World War I, when J. S. Haldane (1860–1936) published his book *The Therapeutic Administration of Oxygen,* in which he detailed his experiences in treating soldiers affected by toxic chlorine gas warfare. Still, methods of oxygen measurement were unavailable. By the early 1920s, physicians tried placing a tube into the patient's mouth, providing a direct path for oxygen to travel and decreasing the amount of gas that escaped. At that time, scientists first suggested that a continuous supply of oxygen be provided. Haldane also developed equipment that could deliver oxygen effectively and relatively cheaply.

As more was learned about the physiological effects of oxygen in respiration, such as a better understanding of the gas exchange process and Otto Warburg's work on the enzyme of cell respiration in 1931 (see Chapter 3), interest in oxygen therapy rapidly grew. In the 1950s, oxygen therapy was further advanced when new methods for safely handling oxygen were developed.

Oxygen Therapy Today

According to the American College of Chest Physicians and the National Heart, Lung and Blood Institute (NHLBI), oxygen therapy is recommended for several conditions, including cardiac and respiratory arrest, as well as systemic hypotension and as "added oxygen" during and after anaesthesia. The therapy is used most widely when hypoxaemia (deficient oxygenation of the blood) is present due to chronic lung and respiratory conditions, including severe pneumonia, chronic COPD, interstitial lung disease, and other respiratory diseases, such as asthma, when they have reached a severe stage.

Physicians now have methods to test and determine the percentage of oxygen in the blood, called *oxygen saturation*. Levels below 90 percent may indicate a need for oxygen therapy. The first test method is called *oximetry*, and uses a clip attached to the ear lobe or several other areas on the body. The second method is more complex and called the *arterial blood gas test*. In this test, blood is drawn from an artery (usually in the wrist) and then the oxygen and carbon dioxide levels are measured.

Three primary types of systems are available to supply oxygen:

- Concentrators—These machines plug into an outlet and take oxygen from the room air.
- Compressed gas—These systems use steel or aluminum cylinder tanks that contain oxygen.
- Liquid—This system includes a large stationary container and a refillable portable unit with a lightweight tank.

The most common way that oxygen is delivered to the body is through a cannula (small plastic nasal prongs) placed under the nostrils. A face mask or oxygen tent is also often used. When long-term continuous therapy is needed in extreme cases, a transtracheal tube, or **catheter**, may be surgically placed into the neck so that oxygen is delivered directly to the trachea.

Although we breathe oxygen every day, oxygen therapy using pure or high concentrations of oxygen should be prescribed just like a drug because of the risks associated with it. In chronic lung disease patients, for example, oxygen may suppress the central nervous system's respiratory functions. Another concern is fibrosis of the lungs. Furthermore, oxygen is extremely flammable.

Overall, however, oxygen therapy has proven to be a safe and often effective approach for treating many severe respiratory problems (see "Breathing for Health"). It has the added benefit in some cases of reducing sleepiness and headaches and increasing mental functioning that may be hindered by low oxygen content in the blood.

MECHANICAL VENTILATION (RESPIRATOR)

Mechanical ventilation is the use of a machine (popularly known as a respirator) to treat acute respiratory failure by inducing inflation and deflation of the lungs and regulating the gas exchange process. The first experiments with mechanical ventilation date back to the seventeenth century when John Mayow (c. 1640–1679) discovered how air was drawn into the lungs by enlarging the thoracic cavity, thus shedding light on the mechanics of breathing. In 1832, Scottish physician John Dalziel (dates unknown) described a respirator-like device that he had used with patients suffering from respiratory stress or failure.

Breathing for Health

Few of us give much thought to breathing. It just comes naturally. However, more and more research is pointing to *how* people breathe as an important aspect of their health. As a result, focusing on a proper breathing pattern or doing breathing exercises may be nearly as important as physical exercise and a proper diet.

Controlled or regulated breathing exercises are nothing new. The practice dates back centuries to Eastern philosophies and exercises like meditation and yoga. *Prana*, for example, is an ancient Sanskrit word that means breath, and *pranayama* literally means "regulating the breath." The practice of pranayama originated more than 5,000 years ago as part of a Vedic culture in what is now known as India. Numerous breathing techniques for health were developed via pranayama, including a technique of breathing through alternating nostrils, which was thought to create balance in the person's overall physiology, improve the function of the nervous system, and benefit many organs.

Over the past three decades, more and more scientific data has been accumulated indicating that breathing exercises are important to health and general overall well being. Many researchers believe that breathing in slower and more deeply is beneficial because it brings more oxygen into the blood. Studies have produced evidence that breathing exercises may impact stress hormone balance and decrease pulse rate, blood pressure, and blood fats such as cholesterol. Some studies have also found that yoga and deep breathing exercises may benefit asthma patients, who are also often instructed to take slow deep breaths when they feel a severe attack coming on in order to keep them calm and breathing as regularly as possible while they seek help.

A study published in the *British Medical Journal* in 2001 found that practices such as yoga and even saying the rosary can help promote slow and deep breathing to positively affect many of the body's vital signs. In the study, researchers told twenty-three test subjects to recite the rosary in Latin or to repeat a typical yoga mantra (the repetition of a word or phrase). The slow deep breathing associated with these practices synchronized the subjects' cardiovascular rhythms, resulting in favorable psychological and physiological effects.

Researchers believe the benefits were not related to the practice of prayer or meditation itself but rather the timing of breaths involved in reciting the rosary or mantra. Rhythmic fluctuations in the circulatory system occur in ten-second cycles, or six times per minute. The researchers found that the yoga and rosary exercises caused people to breathe in precisely the same rhythm, resulting in six breaths per minute as opposed to fourteen for normal, but more shallow, breathing. As a result, say the researchers, the slow deep breathing associated with these practices synchronized the subjects' cardiovascular rhythms, resulting in favorable physiological and psychological effects.

Another study published in the *Lancet* showed that cardiac patients who took twelve to fourteen shallow breaths per minute (as opposed to optimal breaths of six per minute) were more likely to have low levels of blood oxygen, which could adversely affect skeletal muscle and metabolic function and contribute to muscle atrophy and exercise intolerance. In *Aviation, Space, and Environmental Medicine,* study results showed that deep breathing can reduce the symptoms of motion or travel sickness by increasing the activity of the **parasympathetic nervous system** and preventing abnormal rhythm in the stomach. A similar study found that deep breathing can reduce nausea. Researchers believe the effect is due to the closeness of the vomiting and respiratory centers in the brain. When the brain's respiratory center focuses on taking controlled deep breaths, the vomiting center is less able to process thinking about nausea. Some hospital nurses and staff are also being instructed to teach post-operative patients to do deep-breathing exercises to help prevent atelectasis (collapse of an expanded lung) and pneumonia, which are among the leading post-operative complications in patients.

According to some researchers, people have become "shallow breathers," meaning that they are using their diaphragms less and less to breathe and, as a result, only use the narrow top portion of the lung surface for oxygen exchange. Shallow breathers are likely to take a breath and pull in their stomach, which pushes the diaphragm up so the air has nowhere to go. Instead, the shoulders go up to make room. In deep breathing, the diaphragm muscles are lowered by expanding the abdomen or stomach, which elongates the lungs and draws in more air. A good test of deep breathing is to put the palms of the hand against the lower abdomen and blow out all the air. Then take a big breath. If the abdomen expands, you're on the right track.

Research into developing a "machine" that could take over breathing for the lungs continued throughout the nineteenth century. Eugene Joseph Woillez (dates unknown) of France built a mechanical ventilation system in the 1870s. His device is considered the forerunner of the "iron lung," which became a familiar and horrifying sight to the public during the onslaught of polio in the twentieth century. Even master inventor Alexander Graham Bell invented a type of respirator called a "vacuum jacket" in the 1880s.

It was the rise of polio epidemics starting around 1916 that led to the first major advances in mechanical ventilation. Polio results from the virus poliomyelitis. In its extreme form, polio can affect the central nervous system, causing various degrees of paralysis and impair a person's ability to breath by paralyzing the diaphragm and intercostal muscles.

In the 1920s, American scientist Philip Drinker (1894–1972) conducted an experiment with cats and developed a respirator to help polio patients breathe. Drinker enlisted a local tinsmith to make a cylindrical enclosed container with a rubber diaphragm that made the cylinder airtight while a person's head could still stick out. For his test machine, Drinker used a vacuum cleaner blower as the pump to raise and lower pressure within the chamber. The basic concept was simple: each cycle of vacuum within the chamber allowed the person's lungs to be filled with atmospheric air; subsequent increase of pressure forced exhalation of air from the lungs.

In 1928, Drinker conducted a test with an eight-year-old girl who was comatose from lack of oxygen. Within minutes of placing her in the machine she regained consciousness, although she died a few days later of pneumonia. An unknown journalist called the machine the "iron lung."

The iron lung (see photo) became a symbol of fear throughout the United States and the world as polio spread, especially among children. Watching children confined to these huge chambers was devastating. Eventually, Jonas Salk (1914–1995) developed a vaccine for polio in the 1950s, ending the disease's horrific grip on the public.

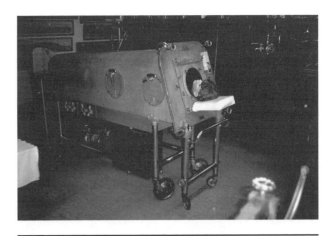

1940s iron lung for polio treatment. © Dell R. Foutz/Visuals Unlimited.

Nevertheless, a number of problems existed with the iron lung. For example, food or vomit could be aspirated into the lungs, and serious skin ulcers could develop in patients who were immobilized for long periods. Still, mechanical ventilation was recognized as a tool that could be applied in many different cases of respiratory failure. Major advances were made in the 1960s as the field of surgery progressed in treating many acute and chronic lung disease patients who would need assistance with breathing following surgery.

Mechanical Ventilation Today

Iron lungs, which are a type of mechanical ventilation known as intermittent negative pressure ventilation (INPV), are still used today, primarily in severe cases involving the brain's respiratory control center or when the diaphragm is paralyzed by spinal cord disease or injury. But even during the polio epidemic, new devices called intermittent positive pressure ventilators were coming into their own and shown to be even more effective than the iron lung in saving lives.

Intermittent positive pressure ventilation essentially works by forcing air into the lungs (and sometimes extra oxygen when necessary) through a face mask, mouthpiece, or nasal mask, and then allowing normal passive exhalation. Invasive mechanical ventilation methods are also used involving a tracheotomy, in which a surgical hole in the windpipe is made and a tube is inserted to provide ventilation.

Modern mechanical ventilation not only takes over the vital role of respiratory muscles to induce rhythmic breathing, it also provides other natural functions of the respiratory system, such as humidification and filtration. The

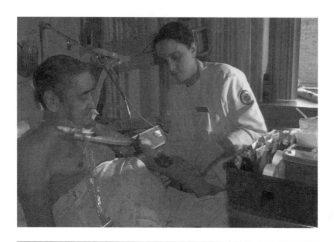

Patient on a ventilator. © Ansell Horn/Phototake.

primary overall goals are to support pulmonary gas exchange, increase lung volume, and unload, or relieve, the ventilatory muscles (diaphragm and intercostal muscles).

There are two phases in the respiratory cycle called high lung volume (inspiration) and lower lung volume (expiration). Prolonging the duration of the higher volume cycle enhances oxygen intake. Modern mechanical ventilators generate and regulate the flow of gas into the lungs until a predetermined volume has been delivered or airway pressure generated. The flow then reverses when the machine moves into the expiratory phase, usually at a preset time or according to a preset tidal volume (amount of air that passes through the lung) measurement. The respirators can either be controlled, in which case the ventilator is active and the patient passive, or assisted, meaning the patient can initiate and may participate in the act of breathing (see "Giving the Patient Neural Control").

Respiratory failure and impaired respiratory function requiring a ventilator occur for a variety of reasons, including lung disease and chest injury. The most common causes are asthma and COPD. Other conditions that cause respiratory failure include pneumonia and pulmonary edema. Mechanical ventilation is also used with newborn infants with severe lung disease.

Although respirators are often used for patients who are severely ill, they have also found a use for milder conditions, such as sleep apnea (see Chapter 4). In this case a continuous positive airway pressure device is used to provide a positive pressure to the inside of the throat and prevent it from collapsing during sleep. In many other cases, mechanical ventilation is used only for a short time until normal breathing can be resumed. However, for chronic or irreversible cases of disease causing respiratory difficulty, mechanical ventilation can be a life-long requirement used either intermittently or twenty-four hours a day.

Modern mechanical ventilators can still cause complications. For example studies in the 1980s confirmed that high peak airway pressures can sometimes lead to further lung damage by overstretching the alveoli. This problem, however, is usually controlled by techniques that limit peak pressures.

Giving the Patient Neural Control

A new type of mechanical ventilator is under development that enables the patient's respiratory center in the brain to control the ventilator. Called *neurally adjusted ventilatory assist*, the device uses a computer to help analyze electrical activity in the diaphragm to adjust ventilatory assistance during a breath and between breaths. The idea is to allow the patient's respiratory center to assume full control of the timing and magnitude of mechanical support, regardless of changes in such things as respiratory drive and respiratory muscle function.

The device works by monitoring electrical signals between the brain and the diaphragm through bipolar electrodes positioned in the esophagus. The signals help monitor the patient's respiration or control the timing and/or levels of ventilatory assistance. As respiratory needs increase and the brain signals the diaphragm for more effort, the neurally controlled system increases the amount of help the ventilator supplies. In essence, the system allows the brain to directly control mechanical support throughout the course of each breath. The new approach may not only improve patient-ventilator interaction and increase comfort, but also reduce some of the risks associated with mechanical ventilation, such as hyperinflation. It is expected to be useful in patients with severe forms of respiratory impairment and in pediatric patients.

RESUSCITATION

The idea of resuscitation, or bringing someone back to life who appears to have died, dates back thousands of years. The ancient Egyptians used an inversion method to help ventilate the lungs and resuscitate drowning victims and others. The method involved hanging an individual by his or her feet and then applying pressure to the chest to aid in expiration and releasing the pressure to help with inspiration. The biblical Old Testament also provides several accounts of what appear to be resuscitation techniques, including the story of Elijah placing his mouth over a child's mouth to bring the child back to life.

The first report of experimental efforts at helping unconscious people to breathe can be traced back to the Muslim physician Avicenna around 1000 CE when he experimented with inserting a tube into the trachea to help with inspiration. In the fifteenth century, Vesalius also experimented with this approach on animals by surgically inserting a tube or reed into the trunk of the trachea and then blowing into it "so that the lung may rise again and the animal take in air." In the 1500s, fireplace bellows were used to blow a mixture of hot air and smoke in a person's mouth. At the time, little was known about the anatomy of the respiratory system, including the now

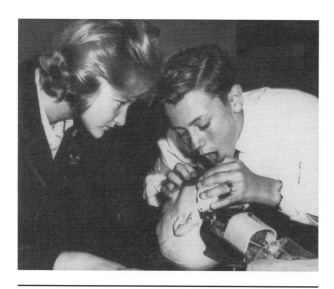

A Red Cross worker demonstrates cardiopulmonary resuscitation techniques to a young woman, c. 1960. © WHO/Red Cross Photo/National Library of Medicine.

known fact that overdistending (enlarging from internal pressure) the lungs by using a bellows could result in death.

Other methods of resuscitation included the eighteenth century method of placing the drowning victim or unconscious individual onto a large wine barrel. The barrel was then rolled back and forth to compress the chest cavity to force air out and then to release pressure to allow the chest to expand and draw air in. Another approach used in the nineteenth century was called the trotting horse method. Lifeguards would place drowning victims onto a horse and run it up and down the beach to get the alternating compression and relaxation of the chest cavity as a result of the body bouncing up and down. Yet another method was rolling a person from side to side sixteen times a minute. To enhance expiration, pressure would also be applied to the back while the victim was lying prone on the stomach.

In the 1950s, mouth-to-mouth resuscitation was established and widely taught to the general public by the American Red Cross. During mouth-to-mouth resuscitation, a person is actually breathing oxygen into the victim's lungs because exhaled breath contains sixteen percent oxygen, which is close to the twenty percent of oxygen contained in the air we normally breathe. This resuscitation technique has been widely used for victims of drowning, electric shock, choking, strangulation, suffocation, some poisonings, and drug overdoses.

Although mouth-to-mouth resuscitation was also used for victims of heart attacks (cardiac arrest), it was far less effective because the heart had stopped beating. As a result, no blood was circulating to carry life-giving oxygen to the brain and other organs. During the 1960s, the modern method of resuscitation known as cardiopulmonary resuscitation (CPR) was developed. CPR combines mouth-to-mouth resuscitation and chest compressions; it has been widely used since 1973 to aid people who have had a cardiac arrest and have stopped breathing. In addition to help with lung ventilation, CPR provides oxygen-rich blood to the brain to help prevent brain damage or death, which can occur in less than eight minutes after a person stops breathing. The oxygen is supplied via mouth-to-mouth respiration, and it is "artificially" circulated to the brain via chest compression. CPR is most effective when it's performed within the first four minutes after cardiac arrest.

Unfortunately, the success rate of CPR on someone who has had a cardiac arrest is only about thirty to fifty percent, and that rate depends on an emergency team arriving within several minutes with a **defibrillator** to shock the heart into pumping again. In these cases, CPR provides a small trickle of oxygenated blood to the brain and heart to keep the organs alive and undamaged until help arrives.

Resuscitation in the Future

Often, even if a person is revived using CPR, the brain can still be damaged because of the lack of oxygen, usually within five minutes of the cardiac arrest. Researchers are currently investigating an approach to prolong this time frame by using a chilled breathable liquid pumped into the lungs.

When the blood stops flowing, the oxygen reserves within cells are completely depleted in about ten seconds, causing a person to lose consciousness. Five minutes later, glucose reserves within the cells also disappear, which begins a process in which cells are literally poisoning themselves with a toxic avalanche of chemical reactions. In the 1980s, CPR pioneer Peter Safar (1924–2003) discovered that this process could be slowed considerably in animals by lowering the body temperature by just 7.2°F.

In 1997, about a decade after Safar's discovery, scientists found that mice could survive immersed in a liquid known as *perfluorocarbon*. The liquid is made up mostly of fluorine and carbon, which are chemically inert and will not harm the delicate alveoli. Furthermore, scientists found they could load perfluorocarbons with gases such as oxygen or carbon dioxide.

Researchers have since developed a device to deliver perfluorocarbon in the lungs using an endotracheal tube, which is usually used to force oxygen into the lungs of an unconscious patient. One goal is for the fluorocarbons to carry enough oxygen to the brain and throughout the body to keep it alive. But the liquid is also chilled so that it cools the blood as it flows through the capillaries in the lungs, ultimately cooling the brain as well, helping to prevent the quick onset of brain cell damage.

Although still in the experimental phase of their research, many scientists are optimistic that "cooling" the body and the brain will help slow down cellular damage due to chemical and inflammatory processes. As a result, more people who are resuscitated will survive with little or no permanent damage to their brain or other organs.

ANTIBIOTICS

Although antibiotics are a common class of drugs that most people with a respiratory tract infection have taken at some time in their lives, their discovery marks one of the great advances that revolutionized medical care in the twentieth century. Since the mid 1940s, antibiotics have saved millions of people with respiratory and other infections by attacking microorganisms that routinely killed people for thousands of years. The word "antibiotic" comes from the Greek words *anti* and *bios*, together meaning "against life."

The history of antibiotics can be traced back several centuries, to when people used natural herbs and molds to treat infections. For example, to treat malarial fever, some South American Indians chewed the bark of the cinchona tree, which contains quinine, an effective treatment for malaria. Some Indians even wore a type of sandal lined with a furry mold to treat foot infections.

The first major scientific advance that would ultimately lead to antibiotics was Louis Pasteur's discovery in 1877 that some microbes inhibited others. This initiated research that, in 1908, led to the synthesis of the sulfa drug prontosil—first of the sulfanilamide drugs—by German chemist and pathologist Gerhard Domagk, who was researching the antimicrobial qualities of several newly developed dyes. This discovery, in turn, led to more studies into various microorganisms that could combat infections. Domagk received the 1939 Nobel Prize in Physiology or Medicine for his work.

The turning point came in 1928 when a Scottish bacteriologist named Alexander Fleming (1881–1955) discovered penicillin. Fleming was working in St. Mary's Hospital in London, England, with the staphylococcus bacteria. He left some in a petri dish (glass plate) and went on vacation. When he returned he found a mold had grown on a jelly-like substance used to feed the bacteria. Fleming quickly noticed that the staphylococcus bacteria remaining in the dish were nowhere near the mold. Looking through his microscope, he deduced that the mold contained something that was killing off the bacteria. He identified the mold, which was similar to the kind that grows on old bread, as *Penicil-*

Penicillin mold growing in culture bottles. Courtesy of the Library of Congress.

lium notatum. Using a filtering process, he discovered the active substance that attacked the bacteria and called it "penicillin."

Amazingly, few paid any attention to Fleming's discovery. According to one account, a university in the United States rejected Fleming's application for $100 to study the drug and nearly fired one of their professors who was going to pay for the research out of his own pocket.

Funding for research did not become available until a decade later when World War II began and thousands of wounded troops were developing infections and infectious diseases. In 1940, Howard Florey (1898–1968) and Ernst Chain (1906–1979) extracted a tiny bit of a yellowish powder from Fleming's mold and found that it protected mice against various bacteria. They then tested the antibiotic on patients and showed that it was effective against many different bacteria, including those that caused blood poisoning. Ultimately, scientists discovered that penicillin-based antibiotics work by inhibiting certain proteins in bacteria cell membranes, causing disintegration of the bacterial cells.

Researchers soon discovered that not all bacterial microorganisms were susceptible to penicillin. As a result, many infectious diseases, such as tuberculosis, did not respond to penicillin treatment. Research began almost immediately for other classes of antibiotics. In 1943, Selman A. Waksman (1888–1973), who coined the term "antibiotics," and his research group found dirt with mold on it from the neck of a chicken. Waksman tested the mold against *tubercle bacilli*, and the mold destroyed it. He called his new antibiotic streptomycin. In 1944, a young woman with advanced pulmonary tuberculosis was being treated at the Mayo Clinic. She had already lost her left lung to the disease and the infection was steadily eating away her right lung. The clinic tested streptomycin on her, and it saved her life.

Over the following years, many other classes of antibiotics were discovered, including cephalosporins, tetracyclines, macrolides like erythromycin, and quinolones. Their effectiveness against a broad spectrum of bacterial diseases and illnesses soon led to widespread use, in-

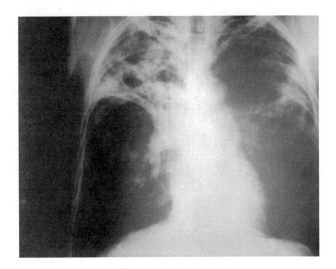

An x-ray of a patient diagnosed with advanced bilateral pulmonary tuberculosis as indicated by the fuzzy white abnormalities on both sides of the lung. © Centers for Disease Control and Prevention.

cluding as a treatment for upper respiratory tract infections, bacterial pneumonia, and influenza.

Antibiotics Today and the Threat of Growing Resistance

The success of penicillin and subsequent antibiotics against a wide range of infections led them to be called the "wonder drugs." Overall, approximately 5,000 antibiotics are known today. About 1,000 of them have been investigated, and only about 100 are used to treat infections. Most antibiotics currently used are produced by actinomycetes, a type of bacteria found in molds and bacteria. The others are produced synthetically.

Bacteria are extremely adaptable, enabling them to evolve rapidly and survive under almost any condition. In fact, within only a few years of Fleming's discovery, scientists became aware that the misuse of antibiotics could lead to resistant forms of bacteria. Fleming himself had experimentally created resistant strains by varying the dosage and conditions of how he added penicillin to bacterial cultures. In 1945, Fleming warned in the *New York Times* that antibiotics could be misused and lead to mutant antibiotic-resistant bacteria.

Nevertheless, it wasn't until the 1970s that the medical community realized that antibiotic resistance was a real threat. The last two-plus decades have seen a major rise in antibiotic resistant infections. The problem had been compounded because scientists found it increasingly difficult to discover new effective antibiotics during the 1970s and 1980s. As a result, newly resistant bacteria rapidly developed while the scientific community found itself falling behind, resulting in about a ten-year gap in the production of new drugs to combat them (see "Developing Resistance to Antibiotics").

Developing Resistance to Antibiotics

When bacteria survive an antibiotic attack they can mutate, or spontaneously change their genetic material or DNA. This change in genetic material results in different kinds of mutations that can fight off antibiotics in different ways. For example, some mutations lead to bacteria being able to produce enzymes that fight off or inactivate antibiotics. Other mutations may prevent the antibiotic from entering into the bacterial cell. Still other mutations may allow the antibiotic to enter the cell, but quickly flush it out again before it can do its work.

Bacteria also acquire resistance by "borrowing" genetic material from other bacteria during a simple mating process called "conjugation," during which genetic material is transferred. Bacteria also can acquire "free" DNA from the environment around them, thus making them resistant. Even viruses can help transport genetic resistant traits among bacteria.

One of the most recent concerns surrounds the rise of drug-resistance in *streptococcus pneumoniae*, which is a common and widespread cause of pneumonia, bronchitis, childhood ear infections, and meningitis. In some parts of the United States, 25 to 35 percent of identified cases are antibiotic resistant. These resistant bacteria have found friendly breeding grounds in hospitals and community homes for the sick and aged. A new resistant strain of tuberculosis has also appeared worldwide, as the disease is enjoying a resurgence.

Scientists have identified several reasons for new strains of bacteria gaining the upper hand over antibiotic treatments. Primarily, most experts point to the physicians overly prescribing antibiotics and patients not using them as prescribed. For example, studies have shown that antibiotics have little or no benefit for most patients with sore throats, maxillary sinusitis, and other upper respiratory infections, as well as those with acute bronchitis. Nevertheless, physicians have continued to prescribe antibiotic treatment, leading to bacteria becoming resistant. Patients also help new strains of resistant bacteria when they don't follow through on their drug regimen. For example, if patients are prescribed a ten-day dose of antibiotics but quit after taking them for only seven days because they feel better, the more resistant bacteria can persist and flourish. The **prophylactic** use of antibiotics over the long term can also lead to resistant strains of bacteria.

Because of the growing threat, the medical community has begun to rethink its use of antibiotics. Family practitioners or primary care physicians will play a crucial role in controlling antibiotic resistance. For example, the inappropriate and frequent use of antibiotics in children with upper respiratory tract infections has been identified as a major contributor to antibiotic-resistant bacteria. As a result, the American Academy of Family Physicians (AAFP) recommends that antibiotics not be used to treat conditions that are unlikely to respond. They also recommend that doctors use narrow-spectrum antibiotics known to fight particular bacterial infections, as opposed to broad-spectrum drugs that cover a wide range of infections.

Antibiotics in the Future

Several new antibiotics have been developed over the past decade, including some that are effective against pneumonias caused by drug-resistant *streptococcus pneumoniae*, as well as vancomycin-resistant organisms. Scientists, government agencies, and drug manufacturers are also working together to develop even newer drugs and approaches to treating respiratory and other infections. For example, some research is focusing on a new class of drugs called *protegrins*, which are based on naturally occurring antibacterial peptides that can be found in a variety of places, from sea-squirts to the white blood cells of pigs to human tears.

Another avenue of research focuses on adapting antibiotics so that they recognize mutant bacteria and destroy them before they reproduce. Some think the ultimate answer may lie with an idea that was first developed before World War I but discarded after antibiotics were shown to be effective. It involves exploiting bacteria's natural enemy, a class of viruses called bacteriophages. These viruses enter bacteria and kill them. Some animal studies have indicated that they may be even more effective than antibiotics in fighting off infections. Yet another promising area of development is based on **genomics** and gene mapping. Scientists are searching for genomes that are necessary for bacteria to survive. Once these genomes are found, it may be possible to develop new antibiotics that target specific genes within bacteria to kill them off.

Although antibiotics are generally taken orally and sometimes injected, the development of inhalation antibiotics is also a burgeoning area of research. It's based on the belief that higher concentrations directly reaching the linings of the lungs will be more potent.

Despite all the research and likely progress in antibiotic development and other new approaches to fight infections, few scientists believe that there will ever be a "magic bullet" that will ultimately solve the problem. Most expect that new resistant strains of bacteria will arise and that the best approach is to get a head start in the race against resistance.

GENE THERAPY

Gene therapy holds the promise of becoming a profound medical breakthrough in the treatment of many diseases, including respiratory diseases, such as cystic fibrosis and lung cancer. The origins of gene therapy date back to the late 1960s when a scientist name Renato Dulbecco (1914–) was studying the ways in which viruses convert normal cells into cancer cells. He found that the viruses accomplish this transformation by introducing their genes into the cells that they infect (see "What Is a Gene?"). Dulbecco

What Is a Gene?

Each human cell has twenty-three pairs of chromosomes, which contain molecules of DNA. DNA contains the instructions that tell various cells in different parts of the body how to function and to react under different circumstances. These instructions are arranged in segments of DNA called genes. Each gene has a code, and these codes ultimately control virtually everything in the body, from growth to digestion to breathing. Genes help the body carry out its various functions, primarily by producing many different kinds of proteins. Each individual has approximately 31,000 genes. These genes contain the coding for 100,000 to 200,000 proteins.

also observed that the viral genes become a permanent part of the infected cell machinery. The fact that the expression of viral genes caused cells to change their properties showed that it might be possible to alter the genetic composition of cells and how they function by introducing new genes into them. The term "gene therapy" was first used in a 1972 article written by Ted Friedman, a pioneering researcher in the field.

Essentially, the gene therapy approach to treating or curing diseases focuses on delivering missing or properly functioning genes to the patient's cells. The hope is that these genes will then be incorporated in the cellular DNA and produce the protein necessary for correct functioning. There are basically two types of gene therapy. Germ-line cell therapy alters reproductive cells (eggs and sperm). Somatic cell therapy alters non-reproductive cells. Germ-line therapy is extremely controversial because it can make genetic changes that could potentially be passed on to the patient's future children. As a result, current medical gene therapy focuses on somatic-cell therapy because it affects only the patient who receives the treatment.

Although the first attempt at gene therapy was conducted in 1990 by National Institutes of Health researchers, the procedure remains highly experimental. One of the primary issues is safety. Scientists have to make sure that the gene will not do anything other than the intended purpose and not produce unwanted side effects.

The research process is also lengthy because scientists must first identify the therapeutic gene and then find a way to deliver it safely to the appropriate cells and DNA segments. The vehicles used to transport and deliver genes to targeted cells are referred to as *vectors*. Researchers are studying many gene delivery systems, including viral vectors, which are viruses made incapable of reproducing themselves, and non-viral vectors, such as "naked" DNA or lipid (fat)-coated DNA. Once the gene is efficiently "delivered," scientists also have to make sure that it "takes," in other words, remains functioning over time within the cell and is not over expressed. On the other hand, a major limitation of current vectors for use in airway gene therapy (where genes are delivered via an aersolized inhalant) is that they have produced only a short duration of the gene's expressions within the cell.

Scientists believe that airway gene therapy is an attractive option for some respiratory diseases because of the potential to easily deliver the genes directly to the lungs. In terms of respiratory diseases, gene therapy research has focused primarily on cystic fibrosis, emphysema, and lung cancer. But other respiratory diseases are also being studied. Researchers in Canada, for example, have used gene therapy to treat rats with asthma.

As scientists identify more and more of the malfunctioning genes that are responsible for diseases, more research will focus on genetic approaches to curing them (see "Completing the Human Genome Project"). For example,

Completing the Human Genome Project

On April 14, 2003, the International Human Genome Sequencing Consortium announced the successful completion of the Human Genome Project, two-and-a-half years ahead of schedule. The international effort to sequence the 3 billion DNA letters in the human genome is considered by many to be one of the most ambitious scientific undertakings of all time, even compared to landing on the moon. The project has produced a human genetic instruction book by producing an essentially complete version of the human genome sequence. This information will be available for scientists to more easily identify genes for combating disease and improving human health. It will also lead to new tools for discovering hereditary contributions to common diseases and new methods for the early detection of disease.

in the case of asthma, scientists have discovered several genes that can play a role in asthma, including a gene known as ADAM33, which was discovered in 2002. Found on chromosome 20, the gene is related to the development of "over-responsive airways" characteristic in asthma. In the May 2003 issue of *Nature Genetics* online, researchers reported on the discovery of another gene that predisposes people to asthma and atopy, a form of hypersensitivity that causes allergies. Called PHF11, the gene comes from chromosome 13 and appears to regulate the blood B cells that produce the allergic antibody immunoglobulin E. Although new treatments for asthma take years to develop, the gene potentially provides a target for drugs that could turn off immunoglobulin E and prevent allergic diseases.

Like asthma, scientists are interested in identifying the genes that may be associated with the development of emphysema. The role of genetics of emphysema is unknown, but scientists do know that only ten to twenty percent of the people at risk for the disease go on to develop full-blown emphysema, indicating that there is a genetic reason for this. In one avenue of research, scientists are looking for "genetic polymorphisms," alternative forms of specific genes suspected to play a role in the development of emphysema, which may differ in a small way among the at-risk population that never gets emphysema. Scientists believe that finding these genes will go a long way toward developing a gene therapy or other new treatments for the disease.

Although promising, gene therapy is still in the very early stages and will require long-term testing on many patients before it can be determined to be a safe and effective therapy. Gene therapy research came under close scrutiny by the Food and Drug Administration and the National Institutes of Health in 1999 when an 18-year-old patient at a medical university died during a gene-therapy experiment. Researchers have determined that the death resulted from an immune response against the vector (an **adenovirus**) used to deliver the gene and not the gene itself. As a result, clinical experiments involving patients are much more difficult to get approval for and are rigorously monitored.

Gene Therapy for Cystic Fibrosis

Although cystic fibrosis (CF) affects multiple organs, it causes the most serious problems through chronic lung disease, which accounts for most of the deaths due to CF. The disease changes the body's normal defense mechanisms that work in the lining of the lung's surface, thus leading to increased susceptibility to bacterial infection. Chronic lung infections in CF patients eventually lead to lung inflammation and damage and infection with *Pseudomonas* bacteria that are impossible to eradicate.

Cystic fibrosis became a primary target for gene therapy researchers beginning in 1989 when the defective gene that causes CF was discovered. One year later, researchers "corrected" CF cells in a lab dish by adding normal copies of the gene. Subsequently, scientists also characterized the gene's protein, called the cystic fibrosis transmembrane conductance regulator (CFTR), which is the molecular cause of CF. In 1993, the first experimental gene therapy treatment was given to an individual with CF. Since that time, more than 200 CF patients have undergone experimental gene therapy treatment.

Cystic fibrosis gene therapy has focused on administering the normal CF gene or protein, as though it were a drug, to the damaged CF airway cells. The goal is to produce a normal protein that will create healthy cells lining the respiratory tract. So far, the approach has achieved some functional restoration of the ion transport defects in respiratory cells affected by CF. But early studies have also reported that acute inflammation of the lung can also occur.

Through the wealth of data gathered so far, researchers are now fine-tuning their investigations into CF gene therapy. For example, they are studying ways to improve the efficiency of the gene transfer by developing vectors that will circumvent the body's complex defense mechanisms within the lungs (see "Developing Inhalation Gene Therapy"). One of the functions of the epithelium lining of the airways is to prevent penetration by foreign materials and invading organisms. As a result, a complex series of epithelial barriers, such as mucous layers, inhibits gene transfer.

Research is also focusing on determining the life span of lung cells affected by CF and identifying the "parent cells" that produce CF cells. As a result, a central question for CF gene therapy is what cell type and what

Developing Inhalation Gene Therapy

Many scientists believe that inserting a therapeutic gene into human cells to fight emphysema, cystic fibrosis, and many other lung disorders may one day be as simple as the patient breathing in a medicated mist. The idea dates back to the late 1980s, when scientists successfully used a mist to place foreign genes in cells lining the lungs of mice. Further experiments revealed that this approach boosted their response to fighting off illness.

In one specific approach to supplying therapeutic genes to the lungs via an aerosolized spray, the genes are attached to circular portions of DNA called *plasmids*. Once the plasmids are inhaled into the windpipe and the lungs, they are carried via liposome vectors that can permeate cell walls. Unlike other gene therapy approaches, this particular method does not seek to incorporate the gene within the chromosome. Rather, the gene remains outside of the chromosome but inside the cell's nucleus where it can begin assembling the needed protein to improve lung health. Because the introduced gene and DNA do not integrate into the host gene or replicate in the body, they only work temporarily and disappear after several weeks. The approach eliminates the risk of damage that could occur when a gene is permanently inserted at the wrong site on a chromosome, or of activating the body's immune response, something that viral vectors have been known to do.

This approach could have broad applications for disease like cystic fibrosis and respiratory distress syndrome. While not curing the problem, the method would be used for ongoing treatments given through periodical inhalation of the gene-containing spray. In the case of respiratory distress syndrome, the spray could help the body fight bacterial infections and then stop after the gene naturally flushes from the body over time.

region (large or small airways) should be targeted. Another issue is related to the repeated application of CF gene therapy. So far, the transferred therapeutic genes' duration of expression is measured only in weeks. Researchers are also focusing on further understanding how CF changes cells that line the airway surface so they can identify the genes and pathways that are altered in cells affected by CF. Other studies are looking at the CF genes involved in making some bacteria in the lungs antibiotic resistant.

Most important, however, is that gene therapy appears to be safe. Many researchers believe gene therapy will ultimately be most effective if given to patients early before the onset of established infection or inflammation in lungs.

Gene Therapy for Lung Cancer

The most common form of cancer diagnosed in the United States, lung cancer accounts for 14 percent of all cancers and 28 percent of all cancer deaths. Since the early 1990s, gene therapy has been studied as a possible innovative approach to lung cancer treatment.

Much of the research has focused on the p53 tumor suppressor gene. It has been estimated that p53 mutations are present in up to 80 percent of lung cancers, particularly in a type of lung cancer called non–small cell lung cancer. The p53 gene mutation is also present in nearly 50 percent of all human cancers.

The p53 gene encodes a protein that responds to damage in a cell's DNA. The p53 protein plays a primary role in suppressing abnormal cell proliferation, or growth, acting like an "emergency break" in the cell cycle. It is also involved in programmed cell death, called **apoptosis,** in which severely damaged cells undergo a programmed death—a type of cell suicide. However, when a cell's p53 genes are lost or mutated, these protective functions cease to work properly. The gene also plays a role in the effectiveness of other types of cancer treatments, such as radiotherapy and chemotherapy, which work in part by triggering cell suicide in response to DNA damage. Response to these therapies is greatly reduced in tumors where the p53 gene has mutated, making the tumors particularly difficult to treat.

In 1997, researchers first tested p53 gene therapy on patients with incurable non–small cell lung cancer. Injecting the gene attached to an adenovirus, they found that the therapy changed the genetic code of lung cancer cells inside the body, the first time that cancer researchers were able to accomplish this feat. Although the therapy did not cure the patients, it represented a first major step: scientists could restore within cancer cells a normal functioning gene that had been defective and caused cancer. Continued investigations showed that the treatment was even more active when combined with a common chemotherapeutic agent called *cisplatin.*

Much of the research into p53 gene therapy for lung cancer involves direct inoculation (injection) of the genes into the tumors using a **bronchoscope** or by **computerized tomography (CT)** scan-directed fine-needle injection. Results from a later study of fifteen patients showed that 80 percent had no active cancer cells at the site treated with gene therapy. With traditional radiotherapy treatment alone, results are typically less than 20 percent.

Focusing on genes and gene therapy in lung cancer has also expanded to other genes. In 2002, researchers identified three specific lung-cancer tumor suppressor genes on chromosome 3. Called 101F6, NPRL2 and FUS1, these genes are believed to be involved in the very earliest genetic changes identified so far in lung cancer, a change that can take place even in normal-looking lung tissue. In addition to the potential for an effective gene therapy for lung cancer, the significance of the finding could be far reaching in terms of earlier detection, diagnosis, and even prevention of the disease (see "Preventing Respiratory Disease with Gene Therapy."). Initial experiments involved introducing the three genes into mice engineered to have human

Preventing Respiratory Disease with Gene Therapy

Although most efforts at gene therapy focus on treating and curing disease, gene therapy also has potential for preventing illnesses in the first place. For example, a gene therapy that targets a gene called FHIT may prove to help prevent Barrett's syndrome (an irritation of the esophagus that can lead to esophageal cancer) and a type of bronchial dysplasia (abnormal growth and development of cells) that occurs in most cigarette smokers and can lead to lung cancer. The FHIT gene causes damaged cells to undergo apoptosis before they can become cancerous and grow out of control. Several carcinogens, including toxins from cigarette smoke, cause this gene to become damaged. Scientists gave a chemical known to cause esophageal cancer to mice that lacked a copy of the FHIT gene. They then gave the FHIT gene to the mice via a viral vector and discovered that the mice receiving the FHIT gene did not develop cancer. While the researchers do not believe that cancerous cells could be cured with the approach, it could help prevent precancerous cells from developing into cancer. In addition, the gene was given to the mice orally, indicating that it may be possible to develop an oral medication based on the gene therapy that could be given to high-risk patients.

lung cancer. The genes dramatically reduced the human lung cancer growth. The therapy also appeared to cause some of the cancer cells to die off. In addition to injecting the genes directly into the tumor to make it regress, the investigators found that a general intravenous infusion of the genes also shrank tumors, even those that had **metastasized**. The study's results were so impressive that the National Institutes of Health gave the scientists permission to start experiments with patients.

Gene Therapy for Rare Cause of Emphysema

Although the majority of people with emphysema developed the disease because of cigarette smoking, people with an inherent defect in the gene that produces the AAT protein have chronic lung inflammation and develop pulmonary emphysema at an early age. The AAT gene limits the activity of inflammatory cells and is crucial to maintaining healthy lungs. The gene also adversely affects the liver and is often fatal. Approximately 100,000 people in the United States suffer from the disease, which is often mistaken for asthma or chronic obstructive pulmonary disease.

In early experiments, scientists first injected the AAT gene attached to a viral vector into the muscle tissue of mice to see if the protein would be secreted into the bloodstream. The method worked and for the first time the

AAT protein was generated at levels high enough to be therapeutic in humans. Furthermore, the results were sustained in the mice for more than four months after a single injection.

In initial experiments with human patients, scientists introduced a liquid containing the normal AAT gene into one nostril of patients with the AAT deficiency. The gene was attached to liposomes (fat globules) used to camouflage the gene so it could get by the body's natural defense systems and enter the epithelial cells lining in the nose. Although less efficient as gene delivery vehicles, liposomes are considered much safer as gene vectors. Upon analyzing the nasal cells and fluid, the researchers found a significant increase in AAT protein in the nostril that was exposed to the gene but not in the other nostril. They also found that the AAT gene had introduced its own DNA in the mix. Investigators are also developing a way to introduce the gene directly to the lungs via an aerosol spray.

PROTEOMICS

Following up on advances in genetic research and the mapping of the human genome, proteomics has also taken a spot on center stage in the study of many diseases and their treatments. The comprehensive study of proteins in human cells, proteomics uses techniques such as **mass spectrometry** and **x-ray diffraction** to determine the precise composition and three-dimensional structure of proteins. The process may also involve comparing protein samples from healthy and diseased individuals. This information can help scientists determine a protein's exact function and gain insights into various physiological processes. By pinpointing disease-associated proteins, scientists are uncovering potential targets for drug development. Proteins may also provide **biomarkers** that can be used to monitor treatment response, select patients who might benefit from a particular drug, and improve disease diagnostics.

Many scientists believe proteomics will play a major role in helping them understand various respiratory diseases and in designing better drugs to treat them. For example, one avenue of research is looking into what role specific proteins play in making bacteria resistant to antibiotics. They have also discovered the structure of a key protein from a **paramyxovirus,** a leading cause of respiratory disease. Paramyxoviruses cause parainfluenza, which leads to illnesses ranging from colds to pneumonia and infects almost all children before they reach the age of five. Scientists have described the three-dimensional structure of a protein (hemagglutinin-neuraminidase) used by paramyxoviruses to latch onto, penetrate, and exit the cells of the respiratory tract. Currently no vaccines or effective drugs are available to stop viruses like those that cause parainfluenza, and the protein discovery may lead to new possibilities for drug development. Scientists have already studied the design of similar proteins to develop effective drugs against other viruses.

Researchers are also looking for novel protein targets for cystic fibrosis. For example, the constituents of mucus and submucosal gland secretions are important areas underlying cystic fibrosis pathology. As a result, scientists are using proteomics to identify how specific proteins are expressed in relation to respiratory tract and lung secretions of people with cystic fibrosis. Once these unique proteins are identified, pharmaceutical researchers may be able to develop drugs to prevent the action or interactions of these proteins, thereby reducing chronic cystic fibrosis and delaying the need for aggressive treatments, such as heart-lung transplants. The proteins may also be used to monitor the disease's progression.

Recently, scientists have also uncovered evidence that emphysema is driven by an excess production and/or secretion of specific lung proteins called *proteases,* types of enzymes that break down the connective tissue of the lungs. These proteases ultimately cause the alveolar sac walls to break down, resulting in emphysema. Some current research focuses on identifying these enzymes and developing techniques to inhibit them as a therapy to prevent further lung damage in patients.

ASTHMA TREATMENTS

The incidence of asthma is increasing throughout the United States and the world. In 2003, the CDC announced that the number of persons with asthma in the United States has increased 75 percent since 1982, while the death rate has increased 40 percent. More than 15 million people in the United States have asthma, and over 5,000 die each year from it. It is the most common chronic disease in children, affecting approximately 4.4 million. Asthma is also becoming an increasingly common cause of hospitalization for children (see photo).

No one knows exactly what causes asthma, but pollution, tobacco smoke, and cockroach droppings are on the long list of triggers that can provoke asthma attacks in people whose lungs are already damaged. Asthma has quickly become one of the major health concerns of the twenty-first century.

Pediatric emergency room: children in the asthma room. © David M. Grossman/Phototake.

Descriptions of asthma date back to 1550 BCE in Egypt when the disease was described in the Ebers papyrus, which is believed to be the oldest preserved medical document. Over the years, people tried many "remedies" to treat asthma, including a mixture of owl's blood and wine and even droppings from camels. Ancient Chinese physicians used the ma huang plant. The plant contains the pharmaceutical component of ephedrine, which was first used in the West to treat asthma in 1924. It is no longer used for asthma but is still a component of many nasal decongestants.

In the late 1800s, Sir William Osler (1849–1919) discovered that asthma was an inflammatory disease of the bronchia. At about the same time, a Victorian doctor named Sir Henry Salter recommended treatment with strong black coffee. Surprisingly, this treatment was not far off the mark. Caffeine is related to effective asthma medications *theophylline*, which is naturally present in tea and is one of the oldest drugs used to treat respiratory disease, and *aminophylline*, which was first used as an asthma medicine in 1908. The forerunner of today's inhaler medications was adrenaline, which was introduced as an asthma medication around 1900.

Modern asthma medications were developed in the second half of the twentieth century as scientists began to delve in the biochemical reactions of cells. New knowledge about the workings of cell receptors, a type of molecular docking station, led to medications that attached to specific cell receptors to implement their therapeutic effects. In 1967, the $beta_2$ receptor was identified as being responsible for bronchodilation, and soon $beta_2$ selective drugs (or $beta_2$ relievers) were developed. A refinement of the steroid hormone cortisone (a type of natural glucocorticoid hormone that works against inflammation) was also achieved in the 1970s, leading to a class of drugs commonly called oral glucocorteroids, or corticosteroids. These steroid-based treatments are used via aerosol inhalers to suppress airway inflammation during asthma attacks. $Beta_2$ agonists to dilate the airways and glucocorteroids to stop inflammation have remained primary treatments for asthma sufferers.

Research into better treatments and possibly even a cure for asthma are ongoing. A major new area of development in asthma research focuses on leukotrienes. Substances called *allergens* enter the lungs and bind to antibodies on inflammatory cells, causing the cells to erupt and unleash a large number of substances called *mediators* to defend against the allergen. Leukotrienes are one class of these mediators and are produced in large quantities by a variety of inflammatory cells in the lungs following an allergic attack. The leukotrienes play an important role in three key features that characterize a severe asthmatic attack: airway swelling, excessive mucus secretion in the airway passages, and constriction of the smooth muscles surrounding the airways.

Several new drugs have been developed that focus on blocking the inflammatory effect of leukotrienes. The drugs hold much promise because they target the underlying disease condition and act by either blocking leukotriene formation or by binding to leukotriene receptors in the airways to prevent swelling, inhibit mucous production, and avoid airway constriction. Although generally well tolerated, leukotriene-targeting drugs have been associated with some side effects, including potential toxicity to the liver and increased risk of infection.

Another avenue of research into new drugs for treating asthma is based on the effects of deep breathing, which is known to open airways after they have closed. The drug methacholine has long been used to study asthma because it narrows airways and causes wheezing in asthmatics but not in healthy people. However, researchers discovered that, when healthy people took only shallow breaths before inhaling the drug, their lungs behaved in a fashion similar to asthmatics and they had difficulty breathing. Further research revealed that healthy volunteers who breathed deeply before taking methacholine did not suffer as much from the drug's adverse affects. Researchers hypothesize that deep breathing may stretch lung tissue, causing the release of a protective chemical that keeps airways open. If they can pinpoint what the chemical is, a new treatment for asthma may be in the works.

LUNG TRANSPLANTATION

Although organ transplantation as a viable therapy for numerous diseases did not occur until the latter part of the twentieth century, the idea of transplanting organs and tissues can be found throughout history. Ancient Hindu mythology, traced to the twelfth century BCE, tells of an elephant's head being transplanted onto the god Ganesha. In the fifth century BCE, myth was already turning into experimentation when Chinese physician Pien Ch'iao was purported to have transplanted hearts between a man with a strong spirit but a weak will and a man with the opposite personality. Ch'iao's goal was to cure the unbalanced equilibrium of the two men's personalities.

Early attempts at transplantation were doomed to failure because of the lack of surgical techniques and in-depth knowledge of human biology. Amazingly, in 1596 Gaspare Tagliacozzi (1545–1599) from Bologna correctly determined that there were many barriers to transplantation. For example, he noted that the "singular character of the individual prevents us from carrying out this procedure on another person." This is a remarkable intuitive understanding made centuries before Peter Medawar (1915–1987) described the principles of immunological rejection and tolerance in the 1950s.

By the beginning of the twentieth century, the first attempts at renal transplantations were made. In 1902, Emerich Ullmann (1861–1937) successfully transplanted kidneys to dogs, using magnesium tubes for blood vessel attachment. Alexis Carrel (1873–1944) together with Charles Guthrie (1880–1963) developed the technique of the sutured blood vessel attachment and successfully performed transplantations of organs and tissues in animals between 1904 and 1920.

In 1954, Joseph Murray (1919–) in Boston, performed the first human renal transplantation with long-term survival of the graft in identical twins. Soon, suppression of the immune system to help the patient accept the donor organ was achieved by whole-body irradiation of the transplant recipients. Pharmacological and biological immunosuppression procedures were greatly advanced in the 1960s, and the importance of tissue classification and tissue matching tests were established.

In 1963, James Hardy performed the first lung transplant on a patient at the University of Mississippi. However, the patient survived for only eighteen days. Over the next decade results remained disappointing, with a median survival of ten days for lung transplant patients. In the 1980s, the first successful heart-lung transplantation and the first successful double-lung transplant were performed.

Overall, lung and most other types of transplants didn't become viable treatment options until surgical techniques improved and better immunosuppressive drugs, specifically cyclosporin, were developed. For a donor organ to be accepted by the patient's body, doctors use immunosuppressive drugs to suppress the patient's immune system and prevent it from rejecting the organ just as it would try to fight off bacterial or viral infections. On the other hand, this suppression leads to transplant patients in general becoming more susceptible to infections and other diseases. Transplanted lungs are especially susceptible to infections because they breathe in air from the atmosphere that may contain bacteria, viruses, or pollutants.

Lung Transplantation Today

Lung transplantation has become a life-preserving option for people with a variety of life-threatening lung diseases. The majority of lung transplantations are performed for four diseases: chronic obstructive pulmonary disease, idiopathic pulmonary fibrosis, pulmonary hypertension, and cystic fibrosis. Lung transplantation is not recommended for patients with lung cancer for several reasons, including the increased chances of cancer reoccurring in transplant patients. By the beginning of the twenty-first century, more than 15,000 lung transplants had been performed worldwide, including 2,700 heart-lung transplants, 6,500 single-lung transplants, and 4,600 double-lung procedures. Survival rates have also increased dramatically over the years, with five-year survival rates reaching the 40 percent and higher range.

Surgeons perform either single- or double-lung transplants depending on the extent of the disease. Heart-lung transplants are often performed when

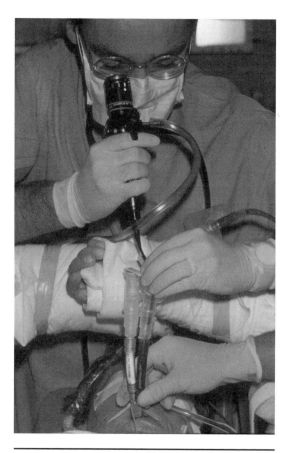

A surgeon uses a bronchoscope during a lung transplant operation. © M. English/Custom Medical Stock Photo.

the patient's heart has been "overworked" or weakened by chronic lung diseases, such as emphysema. Most patients are referred for transplantation only when they are likely to die within a few years from their disease.

Unfortunately, not enough donor organs are available to meet the demand of patients needing transplants. Donor lungs are more difficult to find than many other organs because of their fragile nature; the donor's age at the time of death and the lung's exposure to various harmful substances make them very susceptible to damage. Many factors influence the "health" of a transplanted lung, including air pollutants and cigarette smoke.

Although long-term survival in lung transplant recipients is improving, the survival rate continues to progressively decline in later years, predominately due to the late complication of bronchiolitis obliterans syndrome, a disease of the bronchioles characterized by recurrent wheezing and eventually airway obstruction. Furthermore, the typical lung transplant candidate has a life expectancy of less than eighteen months. Often, lung transplant patients must also depend on supplemental oxygen and have an overall poor quality of life following surgery.

Lung Transplantation in the Future

Despite its many drawbacks, lung transplantation represents the only hope for many patients dying from severe lung diseases. As a result, research continues on improving both the quality of lungs for transplantation and the long-term outcomes for patients. Because lungs are so fragile in nature, they are prone to preservation injury, that is, injury to the tissue that occurs when blood flow is returned to previously cold-stored (blood- and oxygen-starved) lungs. Transplant specialists are conducting research into how to better preserve donor lungs after they are procured. They are especially interested in developing new methods of continuous perfusion (the artificial circulation of blood through tissues) of the organs; the tissue is oxygenated the entire time, so there is less possibility of ischemic reperfusion injury (oxidative damage to the tissue after blood flow is reintroduced).

In a 2003 article in *Nature Biotechnology*, scientists reported on a technique tested in rats that may one day lead to improvements in lung transplantation. Substances known as **free radicals** are often released into the lung and can cause severe damage upon the lungs' removal from the donor. If the damage is too extensive, the recipient's immune system may reject the lungs. In their research, scientists were able to attach an antioxidant, which fights free radicals, onto an antibody that hones in on lung cells, helping to protect extracted rat lungs from free radical–linked damage. Scientists believe that the discovery may one day enable them to provide targeted therapeutics just prior to organ harvest or immediately after to prevent damage due to free radicals.

Many transplant experts believe that the future progress of lung transplantation and improved outcomes will be defined by continued refinement of patient selection criteria. For example, many transplant centers are currently hesitant to provide lung transplants to patients who continue to smoke cigarettes. Furthermore, better treatments for long-term complications and increasing the lung donor pool are also necessary. Living-donor lobar lung donation is also being used increasingly for patients who cannot survive the wait for a cadaveric (deceased) donor. In this procedure, two living donors each contribute a lobe from one of their lungs for transplantation into the patient.

ARTIFICIAL LUNG

The development of an effective and safe artificial lung would have a tremendous impact for nearly 750,000 patients throughout the United States suffering from emphysema, acute respiratory distress, and chest trauma. Nearly 150,000 of these people die each year because of the inability of their lungs to perform the gas exchange process.

The first artificial lung was developed in the early 1950s. Called an *extracorporeal* (outside of the body) *oxygenator*, the machine actually functioned as a combined artificial heart and lung. The artificial heart segment served as a pump to circulate blood while the artificial lung segment performed the gas exchange process as blood passed through it. The early lung substitute was a system in which anticoagulated blood (blood treated so it will not clot) was directly exposed to oxygen as the blood dripped along wires of a vertically mounted metal screen before reentering the patient's body.

However, these early extracorporeal machines had a lethal effect when used for more than a few hours, causing damage to the blood, including an abnormally small number of platelets in the blood (thrombocytopenia), and inability of the blod to clot or coagulate normally (coagulopathy). Such damage can eventually lead to deterioration of organ function. As a result, they were effective for short-term use during operations but were of little use for patients needing support for more than a few hours.

The machines continued to be improved over the years; by the late 1950s, the screen was replaced by a permeable artificial membrane to facilitate the gas exchange process of removing carbon dioxide from and providing oxygen to the blood before it was pumped back into the body. These machines became known as extracorporeal membrane oxygenators (ECMOs). ECMOs have been used worldwide as routine treatment in newborn infants with severe respiratory failure and in adults with respiratory failure. It works well for patients who have had respiratory failure because of infections, such as pneumonia, or trauma, such as smoke inhalation, because they only need to stay on the machine relatively briefly (from a few days to a couple of weeks) until their lungs are healed enough to begin breathing normally.

The long-term use of ECMOs, however, has many problems. People with severe lung diseases such as advanced emphysema, cystic fibrosis, and pulmonary fibrosis are often dying and in need of a transplant. Although ECMOs have been incorporated as a way to keep these patients alive while waiting for a donor organ, these patients generally do not survive on ECMOs long enough for an organ to become available. Problems with ECMOs include the necessity of giving patients a blood-thinning drug (usually Heparin) to keep the blood moving through the machine without clotting. But the blood thinning can lead to internal bleeding, which is especially dangerous if it occurs in the brain. If blood clots do form, they can also cause damage to the brain and other vital organs. Furthermore, the mechanical pumps used in ECMOs damage the red blood cells that carry oxygen.

In the 1990s, as the field of lung transplantation grew, scientists increased their efforts to find an artificial lung. Their goal has been the development of an effective long-term "bridge to transplantation" for patients dying from lung disease while waiting for a suitable donor organ. Several types of experimental artificial lungs have been developed as part of this effort.

In addition to organ transplant patients and patients in intensive care units, the intravenous membrane oxygenator (IMO) is a device intended for patients with acute respiratory distress syndrome, pneumonia, and severe chronic lung disease. The objective is to oxygenate the blood and remove carbon dioxide before it gets to the patient's lungs, with the lungs adding whatever help it can in the process.

The IMO consists of a tube and a pulsating balloon surrounded by approximately 1,000 polypropylene hollow fiber membranes in which gas exchange occurs. The device is technically a catheter with an external power source and gas exchange line. Oxygen is introduced via the external tube and diffuses out through microscopic holes in the fiber membranes into the blood; carbon dioxide diffuses into the fiber membrane and is pneumatically pumped out of the device. The IMO is inserted in a peripheral vein, usually the large femoral vein in the leg, and snaked through the veins to be positioned inside the vena cava system (the major vein system returning blood to the heart). The balloon pulsates (inflates and deflates) at a rate of about 300 beats per minute to move the fibers and stir the blood. This pulsation effectively mixes incoming blood with the oxygen exiting the hollow fiber membranes and oxygenates the venous blood before entering the lungs. Respiration occurs despite severely injured and poorly functioning lungs. As an alternative means of breathing, the device is intended as a temporary substitute for the lungs, giving them time to heal.

Other types of artificial lungs under development would provide support for a longer time and give the patient more freedom to go about their daily lives, untethered to a pumping system. One type is worn on a vest outside the body. This artificial lung also contains thousands of tiny hollow fibers that mimic the lung's anatomy. The device is connected to a synthetic hollow tube implanted under the skin and through the pulmonary artery. The heart's own pumping force drives blood from the pulmonary artery back to the artificial lung and past the hollow fibers, exchanging oxygen from the air with carbon dioxide from the blood through tiny pores. Another tube carries oxygen-enriched blood directly into the heart's left atrium where it is pumped throughout the body, bypassing the natural lungs. Researchers are also working on totally implantable devices that operate in the same manner, using the heart's own pumping power but without requiring any external tubes or devices.

RESEARCH, DISCOVERIES, AND TREATMENTS: A SNAPSHOT

Keeping up with modern medical science and new discoveries about the human body is no easy task. Efforts like mapping the human genome and advanced techniques in microbiology are bringing about a host of new discoveries and advances each year concerning a wide variety of diseases (see "Repairing Lung Damage with Stem Cells"). For example, scientists are not only delving into the basic microbiological structures that make up the respiratory system and malfunction in disease but also into the unique way in which they interrelate. The following is a brief glimpse at some advances and ongoing research concerning the respiratory system and respiratory diseases.

Adult Respiratory Distress Syndrome (ARDS)

Researchers have found that patients with ARDS have damage in an essential protein found in lung surfactant called *specific protein A*. Currently, about 40 percent of patients who develop ARDS die as a result of the dis-

Repairing Lung Damage with Stem Cells

Scientists are making progress in developing new and creative ways to repair lung damage caused by diseases, such as pulmonary hypertension and cystic fibrosis. A group of researchers at the Imperial College of London, for example, have been able to change mouse stem cells into cells needed for gas exchange in the lungs. Stem cells are an unspecialized cell that gives rise to differentiated cells. Stem cells renew themselves for long periods through cell division and, under certain physiologic or experimental conditions, can be induced to become cells with special functions. In their research, the investigators took mouse stem cells and placed them into a specialized growth system that encouraged them to become cells that line the lung where the gas exchange process takes place. The research holds promise for regenerating damaged lung tissue and possibly creating artificially grown lungs for transplantation.

order. Scientists believe that developing drugs to protect specific protein A from damage may help patients.

Asthma

Scientists have found a potential link between food allergies in children and life-threatening asthma attacks that result in respiratory failure and required ventilator support. One study found that more than 50 percent of the children with life-threatening asthma had food allergies (especially to peanuts), compared to only 10 percent in the control group. This finding raises the question of whether life-threatening asthma attacks in children may be triggered by food allergies.

Ongoing studies have also found evidence that the children of a mother who has allergies have a greater risk of developing asthma than children whose mothers do not have allergy problems. This finding is independent of any genetic influence. Specifically, researchers found that children of an allergic mother will go on to develop asthma-like symptoms when exposed to ragweed, whereas the children of a non-allergic mother will not. The results suggest that the allergic mother may impart a unique biological influence during pregnancy and/or nursing, thereby increasing the risk for the development of asthma in her children. Follow-up research in this area is focusing on determining the underlying causes of this maternal influence. The investigators believe that certain immune system proteins produced by the allergic mother during pregnancy and/or nursing can program the immune system of the developing fetus and newborn to respond to allergens in an asthmatic manner.

Bronchitis

Fear over bacteria becoming resistant to drug treatment because of the overuse of antibiotics is especially of concern in patients with chronic diseases, when repeat therapy often leads to a loss of a specific antibiotic's effectiveness. A new antibiotic for battling chronic bronchitis, moxifloxacin HC, has been shown to speed up recovery and lower treatment costs. Compared to patients receiving a standard antibiotic regimen, patients who received the antibiotic required fewer days of treatment and fewer pills, they also recovered faster. Scientists believe that more effective treatments, incorporating shorter courses of treatment and less frequent dosages, will lead to fewer antibiotic resistant microbes. These types of antibiotics will also lead to greater compliance because the overall regimen is easier to follow, which also has an effect on the development of resistant strains of bacteria.

Chronic Obstructive Pulmonary Disease (COPD)

Advances in cellular and molecular immunology have given scientists new ways of looking at lung disease. For example, mice bred to have very specific immune system defects have been shown to develop lung injuries almost identical to the chronic bronchitis and emphysema found in COPD patients. Because of these new research techniques, some researchers are now looking at COPD as a defect or hyper-response of the respiratory system's normal, or healthy, inflammatory reaction to inhaled particles.

An exciting area of research into COPD includes the possibility of lung and alveolar regeneration. Lung regeneration has received a lot of interest in the scientific community due to reports that a substance related to vitamin A called *all-trans retinoic acid* (ATRA) can induce repair of lungs artificially damaged by enzymes. The landmark article showed that the administration of retinoic acid resulted in the regeneration of alveolar tissue in a rat model. ATRA is not a new drug, because it has already been approved by the U.S. Food and Drug Administration as a chemotherapy agent for certain leukemic blood disorders and for solid tumor reduction. Studies have provided the first experimental evidence that ATRA can increase the number of alveoli in the lungs as well as reverse the effects of induced emphysema in laboratory rats. Scientists are conducting research to determine whether ATRA is effective and safe if it is given to humans with emphysema.

In addition to retinoids like ATRA, protease inhibitors are also under investigation for treating COPD. Scientists have pinpointed a class of proteins found in the lungs of COPD patients as being strongly linked to the disease. These particular enzymes break down proteins, especially the proteins that make up the actual structure of the tissue of the lungs; COPD victims appear to have high levels of these proteases.

Cystic Fibrosis

Improvements in antibiotic therapy, clearance of lung secretions, and nutritional support have all played a role in increasing the long-term survival rates of patients with CF. One groundbreaking treatment developed for CF focuses on replacing the DNase enzyme. DNA released from white blood cells that die while fighting bacterial infections is a major factor contributing to mucus viscosity in cystic fibrosis. A naturally occurring enzyme called DNase can cut long DNA molecules into shorter pieces and reduce their stickiness. The enzyme is administered as an aerosol spray. Studies have shown that treatment with the DNase drug can reduce the frequency of severe episodes of lung infection and slightly improve lung function after twenty-four weeks of therapy. Ongoing studies are determining whether the small improvement in lung function seen at six months persists and whether this therapy will retard progressive loss of lung function.

A device called the "flutter" has also been developed to help CF patients loosen the mucus that clogs their airways without the need for conventional chest- and back-clapping therapy. When patients exhale through the handheld device, a special valve causes rapid air pressure fluctuations in the patients' airways. The resulting vibrations dislodge the mucus from the airway walls and promote mucus movement. In a study sponsored by the National Institutes of Health, three times more mucus was cleared with the flutter than after **chest percussion** and vibration by an experienced respiratory therapist or by vigorous voluntary coughing. Treatment with the flutter does not require the assistance of another person, giving the patient more independence. Further studies are underway to determine whether the improved airway clearance may delay the onset of serious lung disease.

Influenza (Flu)

The FDA has approved a number of new drugs to treat the flu. However, the main problem is distinguishing the flu from other illnesses, getting the prescription and the drug in time, and getting appropriate treatment for any complications, such as pneumonia or other bacterial infections. Advances in developing new drugs for the flu have resulted in several drugs that appear to be effective against all known strains of influenza. Scientists are hopeful that these drugs will also work against new strains that may emerge in the future.

In 2003, the FDA has approved a nasally administered flu vaccine that is the first **live virus vaccine** for influenza approved in the United States. The vaccine is approved to prevent illness due to influenza A and B viruses in healthy children and adolescents ages 5 to 17 and in healthy adults ages 18 to 49. It may be especially appealing to younger children and some adults who fear getting shots.

Pneumonia

Because microbial diseases are becoming increasingly resistant to existing drugs, scientists are always looking for new treatments. A group of basic researchers have devised a new approach to developing treatments for *Pneumocystis carinii* pneumonia, which often afflicts people who are immunocompromised. The researchers are targeting the parasite's genetic **messenger RNA (mRNA)** material, which is similar to DNA but a step closer to protein synthesis. They designed a short **nucleic acid** molecule that binds to a section of mRNA in the organism.

The synthetic molecule is called "antisense," and its genetic sequence is biochemically complementary to its target. Because the synthetic molecule takes advantage of the shape as well as the genetic sequence of its target, it binds extremely tightly to the target RNA strand (see Figure 6.1). The syn-

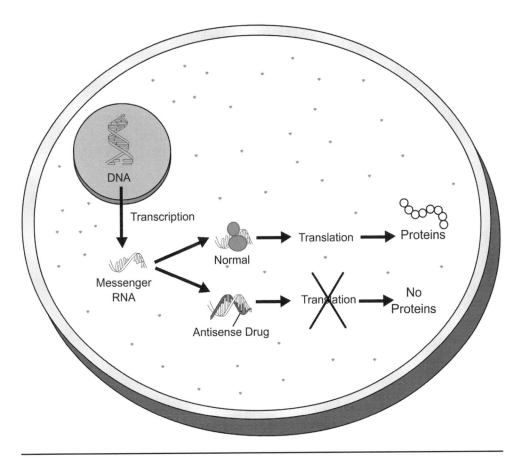

Figure 6.1. Antisense technology.
This diagram illustrates how antisense drugs focus on inhibiting production of specific disease-causing proteins, such as those involved in *Pneumocystis carinii*.

thetic molecule then disrupts the normal process by which *P. carinii* cells construct protein-making machines called *ribosomes*. Without being able to properly construct ribosomes, the organism can no longer grow or reproduce. This particular section of mRNA is an excellent target because it is critical for the parasite's survival but is not present in humans. As a result, drugs that target it are less likely to produce unwanted side effects in patients.

Antisense technology focuses on defeating diseases before the proteins that cause them can form. The production of these faulty proteins begins in the nucleus when DNA forms pre-mRNA, which will leave the nucleus to enter the cytoplasm. In the cytoplasm, this mRNA moves to a ribosome, where it is translated into amino acids. These amino acids are what make up the faulty proteins. The antisense nucleic acid is inserted into the cell's cytoplasm. The DNA of the nucleus encodes the mRNA, which enters the cytoplasm of the cell. But instead of moving to a ribosome to create a protein, the disease-producing mRNA connects to the strand of antisense DNA. So instead of producing proteins, the faulty mRNA is negated. While many traditional drugs focus on the faulty proteins themselves, antisense actually prevents the production of these incorrect proteins.

Pulmonary Hypertension

A relatively rare disease, pulmonary hypertension was virtually untreatable until the 1980s. Pulmonary hypertension gained greater notoriety during the 1990s when a class of popular diet pills became associated with the disease's development and initiated a minor epidemic. Since then, new oral therapies have been developed that hold much promise. One treatment approved by the FDA involves the continuous intravenous infusion of the drug prostacyclin, a naturally occurring molecule.

Endothelin receptor antagonists are also under study as an effective treatment for pulmonary hypertension. Endothelin is a peptide discovered in 1988 and is made in the endothelium (a layer of cells which line the heart and blood vessels). Endothelin constricts blood vessels, elevates blood pressure, and is a potent vasoconstrictor that plays an important role in blood flow. In pulmonary arterial hypertension (PAH), the body produces excess endothelin, contributing to the constriction of blood vessels and affecting the blood pressure in the lungs. Although endothelin is present in healthy people, high concentrations of the substance have been found in the plasma and lungs of patients with PAH, suggesting it is capable of causing PAH or increasing the symptoms of PAH. Endothelin must connect with an endothelin receptor in order to be activated. Endothelin receptor antagonists block endothelin receptors, thereby limiting harmful excess endothelin in the blood vessels.

Pulmonary hypertension is also a life-threatening complication of sickle cell disease. L-arginine is a type of amino acid that helps the body produce nitrous oxide, a potent vasodilator that is deficient during times of sickle cell crisis. This deficiency may play a role in pulmonary hypertension. Nitric oxide therapy by inhalation has improved pulmonary hypertension associated with acute chest syndrome in sickle cell disease, and several studies demonstrate therapeutic benefits of arginine therapy for primary and secondary pulmonary hypertension.

ALTERNATIVE AND COMPLEMENTARY MEDICINE

Therapies are termed as *complementary* when used in addition to conventional treatments and as *alternative* when used instead of conventional treatment. According to some surveys, approximately two out of three Americans use some type of complementary and/or alternative medicine. These therapies may include, but are not limited to, folk medicine, herbal medicine, diet fads, homeopathy, faith healing, new age healing, acupuncture, naturopathy, massage, and aromatherapy. Unconventional and unorthodox, the majority of these therapies are frowned upon by the medical establishment.

Nevertheless, researchers have begun to investigate some alternative approaches to health and treatment in an attempt to see if they are effective and to establish concrete biomedical explanation if they are. For example, researchers are studying the safety and effectiveness of yoga in reducing shortness of breath in people with COPD. Because of some positive research findings in the effectiveness of acupuncture, the prestigious World Health Organization recognizes medical acupuncture as a complementary treatment for a variety of respiratory ailments, including acute sinusitis and rhinitis, the common cold, acute bronchitis, and bronchial asthma (see "Recognizing Acupuncture").

Complementary medicine has even been incorporated into medical school curricula, and some students watch acupuncture demonstrations and attend lectures on Chinese herbal medicine. The University of California at Los Angeles School of Medicine, for example, also has an affiliated Center for East West Medicine. Although students in some medical schools can take alternative medicine electives, such as yoga and mind-body medicine, tried and true medical theories provide the bulk of their coursework.

Alternative therapies, however, can be dangerous. Physicians are especially wary of herbal medications, partially because the FDA does not regulate them, so it is difficult to determine exactly what the patient is getting. Some herbal remedies have been reported to contain a variety of contaminants, from aspirin and steroids to toxic substances such as lead, mercury, and arsenic. Of special concern for respiratory patients has been herbs such as ma huang. The herb contains ephedra, and some asthma and COPD pa-

Recognizing Acupuncture

Acupuncture is one of the few complementary therapies that has a basis in scientific research and has gained acceptance among many physicians as a complement to traditional therapies. Originating in China over 5,000 years ago, acupuncture is based on the philosophy of restoring proper "energy flow" through the body by stimulating "acupoints" across the surface of the body. Overall, there are more than 1,000 identified acupoints in acupuncture. Although most commonly associated with the use of fine needles inserted into different places on the body, these points can also be stimulated via pressure with the hands, electronically, and even with lasers. Research indicates that acupuncture may work by stimulating large sensory fibers to control pain and release pain-fighting chemicals, such as endorphins, throughout the body, triggering the immune system in the process. Many states, including California, issue health care licenses to acupuncturists. The National Institutes of Health endorses acupuncture for various pain syndromes, nausea, and other conditions. The FDA also has approved acupuncture needles as medical devices.

tients have tried it because ephedra is a bronchodilator that can open the airways. But the herb has several side effects associated with heart attacks, strokes, and severe blood-pressure problems. The medical community points out that traditional bronchodilators, such as albuterol, have considerably fewer side effects.

The National Center for Complementary and Alternative Medicine (NCCAM), which is part of the National Institutes of Health, has sponsored several studies of alternative therapies and their effectiveness in respiratory-related diseases. These include studies to:

- determine if osteopathic manipulative treatment of the skeleton and muscles is effective for persons with emphysema as a component of their chronic obstructive pulmonary disease

- assess clinical efficacies and/or adverse effects of dietary borage oil containing gama-linolenic acid and Ginkgo biloba in patients with mild persistent to moderate asthma

- assess the effects of oral magnesium supplements or placebo on clinical markers of asthma control, indirect biomarkers of inflammation, bronchial hyperresponsiveness, and indications of oxidative defense and damage in subjects with mild to moderate persistent asthma. (A number of studies have found an association between low dietary magnesium intake and increased asthma incidence and severity of symptoms.)

For now, anyone considering alternative medical treatments, especially taking herbal and other supplements, to treat respiratory problems should

Regarding Alternative Cold Remedies

The most widely touted alternative approaches to respiratory care have focused on the common cold. Despite some research showing that these alternative treatments may have value, the jury is still out and more research is needed.

Vitamin C, for example, is touted for preventing and curing the common cold. Although these claims are blown out of proportion, an adequate intake of vitamin C is required to keep the immune system functioning properly for fighting infections. Some studies have shown that taking extra vitamin C when a cold first begins may cause a mild antihistamine effect, possibly reducing the intensity of symptoms and shortening the cold's duration. However, the recommended dietary allowance (RDA) for vitamin C is 75 milligrams per day for women and 90 milligrams for men. Megadoses of more than 2,000 milligrams per day may cause side effects such as nausea, cramps, and diarrhea.

Echinacea has become one of the most talked about herbal remedies in the United States, despite the lack of scientific research on the herb by American scientists. A 2001 German study found echinacea alleviated cold symptoms more rapidly than **placebos**. However, another German study published in the *American Journal of Medicine* concluded that treating patients with echinacea did not significantly decrease the incidence, duration, or severity of colds and respiratory infections. Users should be aware that the scientists have not studied the possibility of adverse effects from long-term use of echinacea. Because echinacea is an immune-system stimulant, people with autoimmune diseases like multiple sclerosis and rheumatoid arthritis should not take the herb. It is also not recommended for pregnant or lactating women, or for people taking immunosuppressants.

Another alternative cold remedy are zinc lozenges, which are purported to help reduce the duration and severity of cold symptoms. A 2000 study published in *Annals of Internal Medicine* found that those who took a zinc lozenge every two to three hours during the day had colds that lasted approximately four days compared to eight days in people taking the placebo. However, in a 2000 issue of *Clinical and Infectious Diseases,* other researchers reported that studies showed that zinc lozenges did not reduce the length or severity of colds. Although a zinc deficiency can depress the immune system, taking too much zinc can have the same effect. The RDA for zinc is 12 milligrams per day for adult women and 15 milligrams per day for adult men. Most experts recommend taking no more than 100 milligrams of zinc over the course of a day.

---

consider that little concrete data has been produced to support their effectiveness (see "Regarding Alternative Cold Remedies"). Furthermore the lack of standardized quality control for dietary supplements poses some danger to health, especially in immunocompromised and other severely ill patients. Patients should never take such supplements without first consulting their physicians.

Lung Cancer: An In-Depth Look

Lung cancer is the deadliest of all cancers. As the leading cause of cancer death in the United States, it kills more than 150,000 people in America every year, roughly one American every 3.5 minutes. The incidence of lung cancer is also increasing at one-half a percent each year worldwide—a rapid pace considering the world population of more than six billion people. For now, about 170,000 new cases of lung cancer are diagnosed yearly in the United States and more than one million worldwide. Overall, lung cancer is responsible for approximately 28 percent of all cancer-related deaths, more than colorectal, breast, and prostate cancers combined.

Lung cancer is deadly because it is usually in advanced stages by the time physicians can diagnose the disease. Thus, the number of yearly deaths nearly equals the yearly diagnosis rate. Overall, more than 70 percent of lung tumors are diagnosed only when they have grown very large or metastasized. However, lung cancer is potentially curable when diagnosed early. As a result, a significant body of research is dedicated to developing cost-effective and efficient ways to detect changes in the lungs that lead to cancerous growth.

New cancer therapies are also under constant investigation, and advances in lung cancer treatment over the past thirty years have produced results. Long-term survival for people with lung cancer has nearly doubled over those three decades. Much of this increase in survival has resulted from advances in standard treatments, such as more precisely focused radiation therapy to minimize damage to healthy lung tissue, and combination chemotherapy regimens with fewer side effects. New treatments, such as guided laser surgery have also been developed, and researchers are even working on a vaccine to slow down the course of the disease. Nevertheless, lung cancer remains a highly fatal disease. Although 41 percent of people with lung cancer survive for a year after diagnosis, the five-year survival rate is only 13 percent.

Unlike most other cancers, lung cancer comes with a unique stigma. Nearly 90 percent of all lung cancers result from cigarette smoking and are almost totally preventable. As a result, lung cancer is often associated with personal responsibility or, from another point of view, personal irresponsibility. This viewpoint is undeniably flawed because many diseases can be associated with bad personal habits, including overeating, generally poor dietary habits, and poor hygiene. Blaming the smoker or former smoker who has lung cancer is inappropriate and nonproductive in terms of searching for a treatment and cure.

Approximately 17,000 people (10 percent) who are diagnosed with lung cancer in the United States each year have never smoked. These include people with mesothelioma, a rare form of lung cancer caused by asbestos exposure that affects the linings of the lungs, chest, and abdomen (see photo). Factoring in people who have quit smoking, nearly 20 percent of people who get cancer are nonsmokers. Furthermore, only 10 percent of smokers actually get lung cancer. Because many smokers die at a younger age from other diseases, like COPD and strokes, scientists have yet to make entirely clear the exact genetic and environmental contributions to lung cancer.

Lung cancer has also been found to occur more often in urban areas, with data adjusted for smoking habits. Various industrial processes may be associated with the disease, as well as the long-term effects of diseases such as tuberculosis and rheumatoid arthritis. Even if everyone stopped smoking, approximately 17,000 to 34,000 lung cancer cases still would be diagnosed each year, making it the sixth most common cancer.

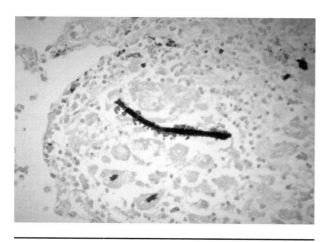

Section of a lung showing a fiber of asbestos or a related substance. © Centers for Disease Control and Prevention.

This chapter expands upon the brief discussion of lung cancer in Chapter 5. It takes an in-depth look at how lung cancer occurs and spreads, the biology of the disease, and current and future treatments. For more on how smoking affects respiration and the lungs, refer to Chapter 8.

LUNG CANCER BIOLOGY

Lung cancer is a disease that occurs when normal lung cells change and begin to divide uncontrollably. Like almost all of the body's cells, lung cells grow, divide, and produce more cells as needed for proper functioning. If cells begin to grow when they are not needed, they form a mass of extra cells that, in turn, form tumors. Benign tumors are not cancerous and usually do not return when removed surgically. Malignant tumors, however, are abnormal and divide without control.

Lung cancer is especially pernicious in that cancerous lung cells spread readily throughout the body. The lungs are richly supplied with blood vessels that normally function as part of the gas exchange process (see Chapter 2). However, these blood vessels are also a convenient route for lung cancer cells to travel and metastasize, that is, become established in other parts of the body (see Figure 7.1). Although most lung cancer cells that enter the bloodstream will die, some do survive and grow.

The lungs are also richly supplied with **lymph vessels**, much like the system of blood vessels. Lymph, or lymphatic fluid, is a transparent, yellow-

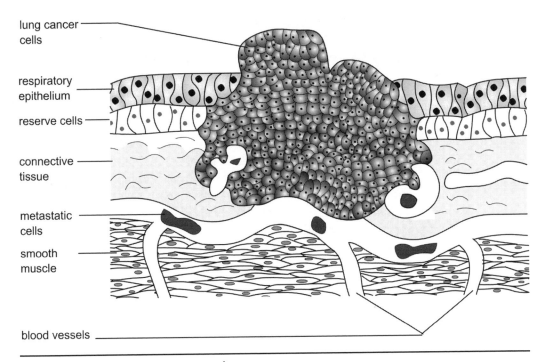

Figure 7.1. Lung cancer metastasis.
This diagram illustrates lung cancer cells entering the blood vessels, where they can travel to other parts of the body and metastasize.

ish liquid derived from tissue fluids. An integral part of the immune system, lymph helps remove bacteria and certain proteins from tissues, transport fat from the small intestine, and supply mature lymphocytes to the blood to help fight infections and other disease. Lymph nodes lie along different points of the **lymphatic system** and help filter foreign invaders, such as germs and cancer cells. In the case of the lungs, there are lymph nodes near the **hilus** (where the large airways and blood vessels enter the lung near the center of the chest) that are usually affected first. Trapped cancer cells can multiply in these lymph nodes, causing them to swell. Enlarged lymph nodes in the neck sometimes indicate lung cancer. Cancer cells often escape the lymph nodes and metastasize.

By far, most lung cancers start in the lining of the bronchi and may be referred to as "bronchogenic cancer." However, lung cancer may also originate in the trachea, bronchioles, or alveoli. Scientists believe that lung cancer is one of the more slowly growing cancers, taking ten, twenty, or thirty years to fully develop into an identifiable cancerous state.

Usually, cells undergo a number of precancerous changes that do not result in a mass or tumor being formed. These precancerous cells often produce no symptoms and cannot be seen by x-ray or other diagnostic techniques. Like other cancers, when lung cells become cancerous, they produce chemicals that cause new blood vessels to form nearby and nourish cancer cells as they grow. By the time the cells are identifiable as cancerous, other cancerous cells have broken away, metastasized, and greatly diminished the chances for a cure.

TYPES OF LUNG CANCER

The two general classifications for lung cancer are small cell and non–small cell lung cancer. The most common lung cancer is non–small cell cancer, which accounts for about 80 percent of lung cancers. This type of lung cancer tends to grow and spread more slowly. Non–small cell lung cancer is usually broken down into three types and named after the kind of cells involved:

- Squamous cell carcinoma—Thin, flat cells that look like fish scales, squamous cells are found in the tissue that forms the surface of the skin, the lining of the hollow organs of the body, and the passages of the respiratory and digestive tracts. Although strongly linked with smoking, squamous cell carcinoma grows more slowly than most smoking-related lung cancers and is often curable with surgery. Approximately 30 to 35 percent of all lung cancers are squamous cell carcinomas and are found near a bronchus.

- Adenocarcinoma—This type of cancer begins in cells that line certain internal organs and that have glandular (secretory) properties. Usually found in the lung's outer region, it is the most common lung cancer in women and accounts for approximately 40 percent of all lung cancers.

- Large cell carcinoma—Lung cancer in which the cells are large and look abnormal when viewed under a microscope. Large cell carcinoma may form in almost any part of the lung. It tends to grow and spread quickly and, as a result, has a poor prognosis. Large cell carcinoma accounts for about 10 to 15 percent of all lung cancers.

Small cell lung cancer (SCLC) accounts for about 20 percent of all lung cancers. It is considered the most aggressive type of lung cancer because it spreads quickly and widely throughout the body. Small cell lung cancer is also known as *oat cell carcinoma* and *small cell undifferentiated carcinoma*. It often begins in the bronchi near the center of the lungs (see Figure 7.2). Almost everyone who gets SCLC has smoked at one time. Nevertheless, there are subtypes of the disease not associated with smoking. Bronchiol-alveolar carcinoma, for example, is most often seen in nonsmoking young females who have a history of systemic sclerosis, lung infections, and autoimmune disorders such as rheumatoid arthritis.

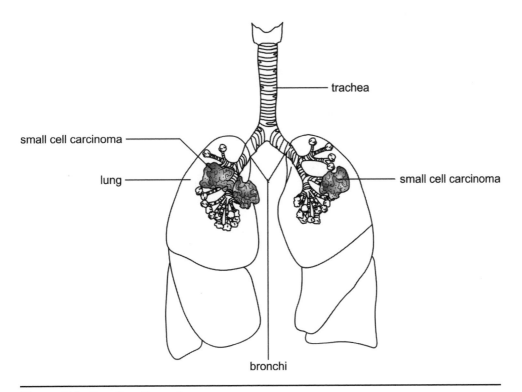

Figure 7.2. Small cell carcinoma.
This type of lung cancer usually originates in the central airways of the lung.

There are many other types of rare lung cancers. For example, atypical carcinoid tumors are cancers that lie between being benign **carcinoids** and small cell lung cancer. Other, even more rare lung tumors include adenoid cystic carcinomas, and various hamartomas, lymphomas, and sarcomas.

CAUSES OF LUNG CANCER

Although smoking is by far the predominant cause of lung cancer, many other substances cause lung cancer in thousands of people each year. Furthermore, researchers are looking into the many biological factors that may make some people more susceptible to cancers caused by smoking or asbestos exposure. For example, only 10 percent of long-term smokers develop lung cancer. The following are the primary causes of lung cancer.

Cigarettes

Smoking is the biggest cause of lung cancer. Studies have revealed that cigarette smoke contains more than 4,000 different chemicals; many of them cause damage to cells and are proven cancer-causing substances, called carcinogens. Developing lung cancer because of smoking is related to:

- Age when the person began smoking
- Number of years the person has smoked
- Number of cigarettes smoked per day
- Depth of inhaling smoke

Secondhand smoke breathed in by smokers also can cause lung cancer and is responsible for an estimated 3,000 lung cancer deaths each year. While lung cancer deaths in men have been falling, it is on the rise in women. Not only are women smoking more, women who smoke also appear to be more susceptible to getting lung cancer than men, possibly because of estrogen or other metabolic factors unique to females.

Racial and socioeconomic background also plays a role in lung cancer rates related to smoking. For example, African Americans have higher rates of lung cancer diagnosis than whites, are diagnosed in later stages, and survive for shorter periods of time. These higher rates can be accounted for partially by socioeconomic factors, such as lack of health insurance to pay for treatment and in differences in care after diagnosis. But researchers have also found that variations in the glutathione S-transferase M1 (GSTM1) gene in some African Americans may cause a reduced capacity to eliminate certain classes of carcinogens found in tobacco.

Researchers have found that ten years after quitting smoking, a person has a one-third to one-half risk rate of developing cancer compared to people who continue to smoke. Furthermore, quitting smoking also reduces the risks of developing heart disease, stroke, emphysema, chronic bronchitis, and other smoking-related illnesses.

Cigars and Pipes

Tobacco smoked via cigars and pipes also increases risk for lung cancer, as well as for cancers of the mouth and others. Risk factors include length of time smoked, depth of smoke inhalation, and number of pipes or cigars per day.

Radon

The second-leading cause of lung cancer in the United States, radon is an invisible, odorless, and tasteless radioactive gas that occurs naturally in soil and rock. Radon not only affects mineworkers but also people in their homes. Radon can come up through the soil and enter cracks in the foundation or insulation and also enter through pipes, drains, and walls. Twelve percent of all lung cancer deaths are attributed to radon exposure, resulting in between 15,000 and 22,000 lung cancer deaths in the United States each year. For someone who smokes, exposure to radon further increases their health risks for lung cancer and other diseases. The Environmental Protection Agency believes that about one of fifteen homes has indoor radon levels at dangerously high levels on a yearly average. Test kits to measure radon levels in homes are available at most hardware stores.

Asbestos

A group of minerals that occur naturally as fibers, asbestos found widespread use in many industries, including the construction and insulation industries. Asbestos fibers break down easily into tiny particles that float in the air and stick to clothes. Inhaled, these particles can lodge in the lungs and damage cells, increasing the risk for lung cancer. Many older buildings may have contained asbestos at one time. By far, most asbestos-related cancers have occurred in workers whose jobs exposed them to large amounts of the substance. **Epidemiological** studies found that workers exposed to high levels of asbestos had a three to four times greater chance of developing lung cancers than workers who were not exposed to it. Once again, smoking further increased the risk in this population.

Pollution

Studies have shown that lung cancer rates are higher in urban areas, and many have linked this increased risk to air pollution and other particulate matter in the air. Researchers have established links between lung cancer and exposures to pollutants, like the by-products of diesel combustion engines and the combustion of other fossil fuels. Environmental pollution may also occur within the home, such as smoke from wood-burning and coal-burning stoves, which has been shown to suppress the immune system. Chinese women and restaurant cooks, for example, are at higher risk because of the vaporized cooking oil they are exposed to when cooking at high heats in a wok.

Many people encounter environmental pollutants via occupational exposure. Occupational exposures are often due to chemical fumes encountered in such industries as roofing, plastic productions, refining, painting, and many others. Other occupations are linked to idiopathic pulmonary fibrosis, which leads to lung cancer. These include farming and livestock handling, hairdressing, raising birds, cutting or polishing stone, and jobs where encountering metal, vegetable, or animal dust is common.

Lung Diseases

Tuberculosis (TB) and certain other lung diseases increase a person's chance of developing lung cancer. For example, cancer can form in the scarred areas of lungs once infected with TB. Young adults who have had Hodgkin's lymphoma, a cancer with a high treatment success rate, are also at greater risk of dying from lung cancer. Someone who has had lung cancer is also more likely to get lung cancer again than someone who never had the disease.

Family History

People with a family history of lung disease are at higher risk of developing lung disease. Scientists are currently researching genetic factors in the disease, such as variances in the GSTM1 gene. This gene encodes for glutathione S-transferase, a group of enzymes that detoxify potentially **cytotoxic agents.**

LUNG CANCER DIAGNOSIS

There are many signs and symptoms of lung cancer, including a persistent cough, chest pain, blood in the sputum, hoarseness, wheezing, and unexplained loss of appetite and/or weight loss. Unfortunately, by the time these symptoms are apparent, the lung cancer has reached an advanced stage and possibly even metastasized.

The most common diagnostic tool for lung cancer is the chest x-ray, which uses low-level radiation to help locate a mass or tumor. Sputum cytology involves looking at individual cells in sputum to characterize them as normal or abnormal. But the test does not specify the size and location of a tumor or tumors.

Other diagnostic tests include **pulmonary function tests** to measure lung function capacities and chest wall mechanics. The test involves breathing through a tube connected to a recording machine. Data analyzed from the test reveal the amount of air the lungs can hold (total lung capacity), how quickly air can be moved in and out (forced expiratory volume), and how well the lungs can transfer oxygen from the air into blood (arterial blood gas). Bronchoscopy involves looking through a tube directly into the inner surface of the major breathing tubes, like the bronchi (see Figure 7.3). In fluorescence bronchoscopy, a dye is injected into the bloodstream and later detected with a special laser to determine if cells have absorbed the dye in a normal manner. If inconsistencies in absorption are found in certain cells, they are removed and tested for cancer, which is called a *biopsy*. Many biopsies are performed through needle aspiration when a needle is inserted through the chest and into the tumor.

In addition to bronchoscopy and needle aspiration, tissue for biopsy is

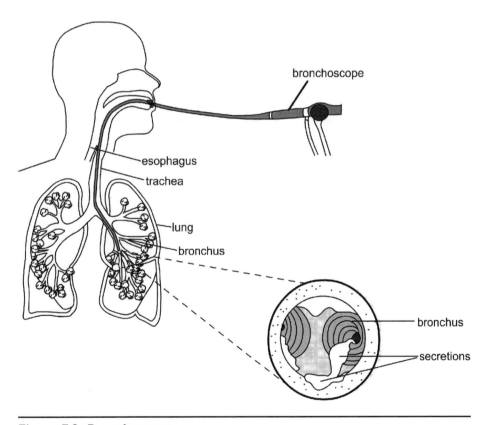

Figure 7.3. Bronchoscope.
This long tubular instrument has a light at the tip and is inserted through the windpipe and bronchial tubes to examine the lung's air passages.

removed via three invasive procedures. A mediastinoscopy involves a small incision in the neck or chest through which a tube and a scope is inserted to observe and obtain tissue samples. This approach is often used to biopsy lymph nodes in the chest to determine whether the cancer has spread to them. In a thoracoscopy, an endoscope is inserted through the skin into the pleural cavity between the chest wall and the lungs. When used with a small video camera, the procedure is called video-assisted thoracoscopic surgery. In a thoractomy, the patient's chest is surgically opened to observe the lungs directly.

Many other tests are also used to further diagnose and stage the disease. Staging lung cancer is simply determining the extent of the disease, particularly whether the cancer has spread to other parts of the body. Staging is important for planning treatments and also because cancer in the early stages is much easier to treat. The stages of non–small cell lung cancer are:

- Stage I—the earliest stage with smaller lung tumors that have not spread to the lymph nodes

- Stage II—a more advanced tumor, or one that has spread to the lymph nodes nearest the tumor

- Stage III—the tumor involves adjacent tissue and/or has spread to more distant lymph nodes

- Stage IV—the tumor has spread to other organs, most commonly the brain, bone, adrenal glands, opposite lung, and liver

Common tests used to stage cancer include the computed tomography (CT) scan, which is a computerized x-ray machine that creates a series of detailed pictures. Magnetic resonance imaging (MRI) uses a powerful magnet linked to a computer to make detailed pictures of inside the body. Radionuclide scanning involves the swallowing or injection of a mildly radioactive substance that is measured and recorded by a scanner to determine levels of radioactivity, which can reveal abnormal areas in certain organs. In positron emission tomography (PET), a signal-emitting tracer is attached to a simple sugar and injected intravenously. The PET scanner records the signals the tracer emits as it travels through the body. Malignant tumors use sugar at a much higher rate than benign tumors or normal tissue. PET scanners can detect this higher usage even in very small malignant tumors.

LUNG CANCER SCREENING

Although screening for early detection is a common approach for many cancers, no major medical organization in the United States supports screening for lung cancer. The idea behind screening for cancers is to catch the disease in the earliest stages, which improves treatment outcomes, primarily for survival rates. Cancer screenings have been effective in improving outcomes for breast (mammography), prostate (PSA test), colorectal (colonoscopy), and cervical (pap smears) cancers.

In the mid-1970s the National Cancer Institute (NCI) conducted studies using chest radiology and sputum cytology testing to identify persons with lung cancer. The results of these studies failed to show any benefit to cancer patients in terms of survival outcome. Both chest x-ray (radiograph) and sputum examination (cytology) were declared to be insufficiently accurate in routine screenings of asymptomatic people, and mortality outcomes remained largely the same for patients who were screened and those who were not. Furthermore, repeated chest x-ray screening is considered to be potentially harmful due to cumulative radiation exposures.

Nevertheless, early stage I lung cancer has a cure rate of 70 percent, and the debate over lung cancer screening has once again come to the forefront. Some scientists claim that the early NCI studies were flawed in their design, execution, and analyses, making the findings questionable. Other radiologists believe that, while chest x-rays have their limitations in terms of early detection, chest x-ray technology has improved over the years. As a result, hundreds of thousands of lives would be saved each year if screenings targeted high-risk populations. Improved imaging technologies and better molecular detection of cancer markers in expectorated sputum have also helped to improve early diagnosis capabilities.

A new area of development is spiral CT scanning for lung cancer screening, which is much more sensitive and accurate than x-rays or sputum cytology. Spiral CT scanning uses x-rays to scan the entire chest in about fifteen to twenty-five seconds. During this time, the CT scanner rotates around the patient on the table, gathering a series of very narrow x-rays taken at varying depths of tissue. A computer creates images from the x-ray information and assembles them into a three-dimensional model of the lungs.

In a study of 1,000 smokers, researchers in the Early Lung Cancer Action Program (ECLAP) study found that annual screenings using low-radiation-dose spiral CT scanning successfully detected very small cancer lesions that were often not visible via x-rays. Overall, CT scanning detected thirty-one tumors not found with x-rays, and nearly 85 percent of these tumors were found to be in the earliest stage of development. Studies around the world have duplicated ECLAP findings that more than 80 percent of stage I lung cancer cases can be detected through CT screening as compared to current early detection rates of about 15 percent. Proponents of CT scanning for lung cancer point out that people who have stage I lung cancer that goes undiagnosed will probably die within five years. Conversely, people with stage I lung cancer who are diagnosed and undergo surgery have an 80 percent survival rate at five years.

It is now well established that CT is far more specific than conventional x-rays in diagnosing early peripheral lung cancers, which are usually adenocarcinomas. In Japan, CT is a standard detection approach for lung cancer, and mobile scanners are used to screen for the disease in rural and other populations.

Despite these findings, no long-term studies in the United States have proven that CT screening for lung cancer leads to better outcomes and actually saves lives. Organizations like the American College of Radiology are not likely to recommend CT screening for lung cancer until more data is available. Nevertheless, many scientists are beginning to call for CT screening as an approach for finding early-stage lung cancer in especially high-risk populations, such as people over 60 years of age who smoke.

One of the major issues concerning lung cancer screening is the associated costs in relation to beneficial outcomes. To make the screenings cost effective, many in the medical field say it is essential to establish strict criteria for who should be screened. Some argue that low-dose CT screening, which cost about $300 per person in 2002, is cheaper than standard CT scanning, and only slightly more expensive than chest x-rays. They point out that as the demand for CT screening increases the cost will decrease, just as it did for mammography screening for breast cancer. Furthermore, it is less costly to treat cancers in the earlier than later stages. Another important factor is that researchers have found that false positives (identifying a tumor as cancerous when it is not) are uncommon with repeat screenings.

The issue of lung cancer screening remains complex for a variety of reasons. For example, false positives can cause people anxiety, fear, and lead to rising health care costs. Furthermore, overdiagnosis of the disease can lead to unnecessary treatments for tumors that may never become life threatening or require medical treatments.

Medical science is currently in the midst of a technological revolution that will ultimately change how lung cancer is identified and treated. Over the course of the ECLAP study alone, CT technology has continued to improve. When the study began in 1992, researchers were using a single-slice helical CT scan that acquired thirty 0.4 inch (11 millimeter) slices to image the entire chest. By 2003, CT technology could acquire 0.03 inch (1 millimeter) slices and generate 300 detailed images. Further improvements in CT scanning will likely include software for automatic reading of scans to determine whether "nodules" have changed over a period of a few months.

Few believe that everyone will quit smoking, which would prevent an estimated 90 percent of lung cancer cases from ever happening. And because of the high impact in cost of lives and costs to society, lung cancer screening will likely become a reality one day. A major challenge is to determine exactly who is at high risk and should be screened. Then image results must be linked to an overall plan to detect and manage the disease.

LUNG CANCER TREATMENT

Scientists have made progress in developing treatments for lung cancer, and knowledge about the disease and how to treat it is increasing. Nevertheless, lung cancer kills more people than any other form of cancer, and the overall 13 percent, five-year survival rate is relatively low.

Current treatments for lung cancer fall into three categories: surgery, radiotherapy, and chemotherapy. These treatments are seldom used alone but rather are combined in various forms for a cumulative treatment effect. A type of laser therapy called *photodynamic therapy* is also growing in popularity for use in certain early-stage non–small cell lung cancers.

Surgery

Surgery focuses on removing the cancer from the lungs and works best in the early stages before the cancer has metastasized. As a result, it is the treatment of choice for non–small cell cancers that are still small and have not spread. The type of surgery depends on where the cancer is located in the lungs. The three major types of lung cancer surgeries are:

- Segmental or wedge resection, which removes only a small part of the lung
- Lobectomy, which removes a lobe of the lung
- Pneumonectomy, which removes one of the lungs (individuals can often breathe normally with only one lung)

Surgery is usually not the first line of treatment if the cancer has spread to the other lung or other vital organs, or if the cancer is in a difficult place to reach. It is rarely used in small cell lung cancer because this type of cancer has usually metastasized before being diagnosed. The surgical side effects vary in severity and longevity. Lasting pain in the incision sites and stress are the most common complaints. As with all surgeries, there are chances for infection and other surgical-related problems.

Radiation Therapy

Radiation therapy, or radiotherapy, is the use of high-energy x-rays to kill cancer cells. The goal is to focus the radiotherapy on the cancer cells while doing as little damage as possible to normal cells. Radiation therapy is often used in combination with surgery and chemotherapy. For example, it can be used before surgery to shrink a tumor for removal or after surgery to destroy cancer cells that may remain in the treated area. It is also combined readily with chemotherapy treatments in cases where surgery is not appropriate.

Radiotherapy is often used as a main treatment in non–small cell lung cancer, especially if the cancer has not spread and cannot be removed through surgery. It is also noted for its ability to lessen specific problems and symptoms, such as pain, blocked airways, and shortness of breath. It is also recommended as an addition to chemotherapy patients who have improved results. In this case, radiation therapy may help prevent the cancer from spreading beyond the chest.

Radiotherapy can be applied externally or internally. External radiotherapy involves the aiming of x-rays from outside of the chest onto the lung. By killing growing cancer cells, radiotherapy helps stop or slow down the cancer cells from multiplying. Timing and dosage are major concerns in radiation therapy, and much research is still being done in this area. For example, CHART (continuous hyperfractionated accelerated radiotherapy) represents a daily dose of radiotherapy divided into more than one treatment each day, sometimes in fractional amounts. CHART patients usually receive such treatment seven days a week until the prescribed course is finished.

Internal radiation involves the implant of a small container of radioactive material placed directly into or near the tumor. Endobronchial radiotherapy, or brachytherapy, is a form of internal radiation used when a tumor is blocking an airway and has made the lung collapse. It involves inserting a thin tube into the lung through which the radiation is introduced for a short period.

Radiation therapy's side effects may include nausea and vomiting, loss of appetite, sore throat and difficulty swallowing, skin irritations, and fibrosis of the lungs and surrounding tissue. Anti-emetics (seasickness drugs) are often used to treat some of these side effects. Radiation therapy, however, does not make a person radioactive and unsafe to be around.

Photodynamic Therapy

Photodynamic therapy (PDT) has been approved by the FDA as an alternative therapy for certain non–small cell lung cancers found in the **tracheobronchial** airways when surgery and chemotherapy are not effective. Photodynamic therapy is also used to reduce some symptomatic problems associated with certain lung cancers, including cancerous tumors that grow in the large airways and cause pneumonia, shortness of breath, or coughing of blood. Photodynamic therapy is also known as *photoradiation therapy, photochemotherapy,* and *phototherapy.*

Photodynamic therapy begins with a photosensitizing chemical injected into the bloodstream. This chemical is then absorbed by cells throughout the body. It rapidly leaves normal cells but remains in cancer cells for longer periods. Usually within forty to fifty hours after the injection, a bronchoscopy is performed. A low-power red laser light is then used on the cancer through the bronchoscope. This light can penetrate one centimeter into the treated area. The laser light is absorbed by the photosensitizing chemical in the cells, producing a **photochemical reaction** that kills the cancer cells. Timing is key to PDT treatment; the laser treatment must be conducted after most of the chemical has left normal cells but before it leaves the cancerous cells. The treatment usually is repeated after two days to remove the dead cancerous tissues, as well as mucus and debris that may have accumulated in the air passages.

PDT has many advantages. Unlike the serious side effects that place certain limits on the use of radiation and chemotherapy treatments, PTD has few side effects. As a result, physicians can repeatedly treat tumors and control their growth when a cure is not possible. While studies are underway for developing implanted optical fibers to deliver PDT to deep-seated tumors within the body, current technology requires that the tumors be in the tracheobronchial airways.

PDT does have some side effects. The most notable is that patients often experience photosensitivity (sensitivity to light) for several weeks. Photodynamic therapy also produces some pain because the tissue break down causes inflammation. Mild shortness of breath may also occur.

Chemotherapy

Although the use of cytotoxic drugs to kill cancer cells is widely known as chemotherapy, the term *chemotherapy* was once applied to all drugs used to treat certain diseases. In this sense of the word, chemotherapy dates back centuries to such examples as the Indians of Peru, who used cinchona bark to treat fevers and malaria.

The modern age of chemotherapy began with the German scientist Paul Ehrlich (1854–1915) (see photo), who developed an arsenic compound called *arsphenamine* in 1909 to treat syphilis. Erhlich's basic ideas for developing drugs holds true for chemotherapy today. He proposed that the chemical makeup of drugs must be studied in relation to how they acted and their affinity for specific cells. The goal was to attack the disease while not harming the patient's healthy cells.

Portrait of Paul Ehrlich, c. 1900–1920. Courtesy of the Library of Congress.

Most early chemotherapeutic agents focused on infectious diseases. In the 1930s, surgery and radiation still remained the only treatment for cancers of any kind. The first chemotherapeutic agent for cancer came during World War II.

When a U.S. Navy ship exploded and sank in Italy on December 2, 1943, its cargo of mustard gas exploded as well. Mustard gas is lethal and was used in battle during World War I. Doctors examined the fatal casualties who had been exposed to the gas and found that their bone marrow cells had disappeared and their lymphatic system greatly atrophied. In 1942, Alfred Gilman (1908–1984) and Fred Philips (dates unknown) tested the gas on mice and then a person with lymphoma. They found that lymphoma cancers experienced a reduction in size when exposed to mustard gas. In 1946, researchers discovered that nitrogen mustards, which differed from mustard gas only because they have an added nitrogen atom and not a sulphur atom, reduced tumor growth in mice. The approach went on to become part of modern chemotherapy treatment for lymph gland cancers, such as Hodgkin's disease.

Over the years, a number of chemotherapeutic agents have been developed. Most of them work by altering or attacking DNA synthesis or function within a cell. They include alkylating agents, antimetabolites, plant (vinca) alkaloids, antitumor antibiotics, and steroid hormones.

Much like Erhlich predicted, these agents work primarily by focusing on changing the cancer cells while sparing healthy cells. They do this by interfering specifically with the cell division cycle of cancers. Antitumor antibiotics, for example, prevent cell division by binding to DNA and making it unable to separate, or by inhibiting RNA and preventing enzyme synthesis.

Typically chemotherapy is given via an injection into the blood stream, or with a catheter inserted into a major vein; some are taken orally. Studies are also underway on inhaled chemotherapy agents that would directly enter the lungs and result in high drug levels in lung tumors, as compared to intravenous chemotherapy. Many chemotherapeutic agents have been developed over time, and modern cancer chemotherapy is most often a combination of various agents designed to treat specific cancers.

In terms of lung cancer, chemotherapy began to be used in combination with radiation therapy in the 1970s to help prevent lung cancer from spreading and to sensitize the tumor so it could be killed more effectively by radiation. Since that time, much effort has been focused on determining the proper combinations of chemotherapy and radiation therapy in terms of dosages and length of administration.

Chemotherapy has been especially effective in shrinking non–small cell lung cancers in some people, thus controlling symptoms and prolonging a better quality life. Lung cancer drugs are also used before surgery and radiotherapy to improve results and after surgery and radiotherapy to help prevent the return of cancer. Chemotherapy is thought to work best in younger cancers because the cells are dividing rapidly, unlike older and larger tumors that slow down and stabilize in growth.

The injected forms of chemotherapy usually consist of a regimen that lasts only a few days, followed by a long period of rest by the patient so he or she can recover from the drugs' side effects. Although many people experience few side effects from chemotherapy, they can be very unpleasant, such as nausea and vomiting. Chemotherapy can also cause mouth ulcers, fatigue, and hair loss, all instances where cells divide rapidly. Because chemotherapy is not site specific, it also affects many of the body's normal cells, including blood and skin cells and cells lining the intestines. This widespread effect can cause anemia and the immune system to become depressed.

FUTURE LUNG CANCER TREATMENTS

Engaged in a fierce battle with a deadly disease, researchers are continuing to look for new ways to better treat—and one day cure—lung cancer patients. For example, radiation therapy is being refined through a technique called *gating*. One of the problems associated with radiation therapy of the lungs is focusing in on the tumor and sparing surrounding tissue, which is made difficult because of lung movement while delivering the radiotherapy. In gating, a computerized x-ray follows the movement of a marker positioned on the patient. Information fed back to the computer causes the radiation therapy beam to be fired only when it is locked in on the tumor area. The result is that more radiation can be given to the tumor or cancer while sparing more of the healthy tissue.

The following are some new approaches and ideas that may help improve lung cancer survival rates until a cure can be found. Although some of the treatments are new specific therapies that stand alone, many others are proving to be effective adjuncts, that is, helping other therapies work better. In fact, cancer is such a complex disease that any treatment or potential cure will most likely be a combination of established, traditional treatments and new therapeutic approaches.

Growth Factor and Growth Factor Receptors

Many individual chemotherapies are approved by the FDA for use in lung cancer. Most current research into these substances focuses on administering various combinations of the drugs, which include cisplatin, carboplatin, paclitaxel, topotecan, ifosfamide, and many more.

Current chemotherapies, however, are not specific to lung cancer cells, and 85 percent of lung cancer patients still die within one year of diagnosis. New knowledge about the molecular and cellular biology of lung cancer over the past two decades has led researchers to take different approaches. For example, researchers now know more about the differences between lung cancer cells and normal bronchial epithelial cells, which is providing new targets for cancer drugs to work more effectively.

A large body of research has focused on growth factors and growth factor receptors. Growth factors are proteins that help cells grow and survive. Growth factor receptors are the proteins that are activated, or turned on, when they attach to growth factors and help promote rapid cell growth. One area of research focuses on epidermal growth factor receptor (EGFR), which is found on the surface of many types of cancer cells. Epidermal growth factor receptors allow epidermal growth factor, a protein, to attach to them. The docking leads to a chemical called tyrosine kinase, which triggers chemical processes inside the cell to promote cell division and growth.

Several studies are looking at substances to block the EGFR, which is overexpressed on malignant cells, including 40 to 80 percent of non–small cell lung cancers. In May of 2003, the FDA approved Iressa (gefitinib), an EGFR inhibitor, for use in advanced non–small cell lung cancer. Iressa was reviewed and approved under FDA's accelerated approval program, which allows patients suffering from serious or life-threatening diseases earlier access to promising new drugs.

Iressa works by inhibiting the activation of tyrosine kinase and switches off the signals from the EGFR that eventually lead the cell to grow and divide. As a result, it is called a *signal transduction inhibitor*. Some scientists believe the drug's greatest benefit may be as a maintenance therapy following chemotherapy. By helping to keep cancer cells dormant for a longer period of time, ultimately Iressa and other similar drugs may help scientists toward their goal of improving lung cancer to a manageable chronic disease.

Angiogenesis Inhibitors

Another new approach in treating lung cancer is to use drugs that prevent the development of new blood vessels. These blood vessels nourish cancerous growth with oxygen and nutrients. **Angiogenesis inhibitors**, such as thalidomide, are being tested to treat lung cancer. Preventing angiogenesis is critical because, once new blood vessels enter the tumor, several scenarios are possible:

- tumor rapidly expands
- tumor invades local organs
- cancer cells escape through new blood vessels and circulate to other parts of the body
- escaped cancer cells metastasize and lodge in other organs

Antiangiogenic therapy specifically attacks growing new blood vessel cells, which occurs only in cancer cells and not in normal healthy tissues. Most antiangiogenic drugs do not kill cancer cells directly and have fewer and less severe side effects than chemotherapy. To keep cancers from regrowing, patients may need to take antiangiogenic drugs as a chronic therapy for the rest of their lives.

COX-2 Inhibitors

Some scientists are calling COX-2 inhibitors a drug developer's dream. In the COX family of enzymes, COX-1 (cyclooxygenase 1) is responsible for maintaining the health of the lining of the stomach and of blood flow to the kidney. However, COX-2 (cyclooxygenase 2) is present in various inflammations, such as arthritis. As a result, COX-2 inhibitors have been developed to treat various inflammatory problems. In 1999, they were also approved by the FDA to treat a type of colon cancer called *familial adenomatous polyposis* (FAP).

COX-2 turns out to be present in many types of cancer cells but especially in cancer of the lung, where it occurs in 90 percent of all cases. Scientists are excited about the therapeutic potential of COX-2 inhibitors for lung and other cancers because, unlike current chemotherapeutic agents, COX-2 inhibitors are very specific. They are highly expressed in precancer or cancer cells and not in others. Furthermore, blocking COX-2 enzyme is not toxic and does not disrupt normal cell function. The effects of COX-2 inhibitors on cells are measurable and can help determine its clinical benefit. Finally, COX-2 appears early in many cancers and blocking its effects may help prevent cancer.

Scientists believe that COX-2 inhibitors may also enhance the effectiveness of chemotherapy. In one study of adding COX-2 inhibitors to a chemotherapy regimen, the investigators found that 28 to 30 percent of patients had tumors that were no longer present or dead, or only a few cells remaining, compared to the normally three to five percent of tumors that would be completely dead at the time of surgery after a regimen of chemotherapy. Scientists also know that COX-2 inhibitors improve the rate at which cancer cells die following a dose of radiation therapy.

Radiofrequency Ablation

An emerging treatment for lung cancer is radiofrequency ablation. Primarily used in conjunction with surgery and/or chemotherapy, radiofrequency involves inserting a needle through the skin and into the tumor, much the same as a standard biopsy. A radiofrequency is then sent through the needle to heat and destroy the tumor. Radiofrequency generates its heat by producing a high-frequency electrical current that causes frictional movement of ions.

As a minimally invasive procedure, radiofrequency ablation offers many advantages. It can be used in patients with multiple cancers who could not withstand extended surgical and chemotherapy treatments. It is also valuable for treating multiple tumors of the lung. The radiofrequency treatment usually lasts between fifteen minutes to an hour, and the patient usually can go home from the hospital the next day. Complications from the procedure have been minimal.

Immune Response Stimulation and Vaccines

Immune-based therapies are appealing cancer treatments because they can target and attack cancer cells while avoiding damage to healthy tissue. Many immune-based therapies are under study for a variety of cancers. For example, researchers have shown that a protein called *secondary lymphoid tissue* (SLC) significantly inhibits lung cancer growth by stimulating an immune response against the disease. Secondary lymphoid tissue is found in the lymph nodes and is a member of the protein group called *chemokines*. Chemokines attract lymphocytes and dendritic cells, both of which are central to proper immune system functioning. When injected into tumors in mice, researchers found that SLC attracted lymphocytes and dendritic cells to the tumor, where these cells then attacked the cancer cells.

Dendritic cells are very successful in fighting off diseases but are unsuccessful when it comes to cancer. Scientists believe that cancer can trick the immune system and prevent the dendritic cells from responding to tumors. Researchers have found that SLC appears to significantly increase immune-enhancing proteins while decreasing the number of immune-inhibiting proteins in tumors. This effect could help prevent important immune cells, such as dendritic cells, from being tricked into ignoring invading cancer cells.

In one study, SLC eradicated lung cancer in 40 percent of laboratory models and dramatically slowed tumor growth in the remaining mice. In addition to its role in stimulating the immune system, SLC has also been found to limit blood vessel formation, which also helps limit cancer growth.

Boosting the immune system lies at the center of the effort to develop a vaccine for lung cancer. Unlike other vaccines, such as those for pneumonia or tuberculosis, lung cancer vaccines are designed to be therapeutic rather than preventive. While they may not prevent the disease, they do focus on helping the patient live longer by prolonging cancer remission periods after treatment.

In one approach, researchers are essentially making patient-specific vaccines by using the patient's own tumor tissue to manufacture the vaccine, known as an autologous (taken from an individual's own tissues, cells, or DNA) tumor vaccine. Immune system cells (such as lymphocytes) are drawn from the patient's own blood, duplicated in the laboratory, and then mixed with certain proteins taken from the lung cancer cells. The mixture is irradiated and then injected into the patients. Researchers have found that the process stimulates an immune response against cancer cells.

Although still in the early stages, research in this area shows that the immune response against lung cancer can potentially be enhanced through a vaccine strategy. This approach has been hindered by the hours and hours of lab work needed to modify tumor cells for an individualized vaccine. Nevertheless, the research opens the door for many more studies to develop this approach into an effective treatment.

EARLY IDENTIFICATION AND PREVENTION

Getting people to quit smoking is by far the best attack against lung cancer. Although treatments have been successful and there is much hope for future improvement, lung cancer largely remains unmanageable. As a result, scientists view early diagnosis and prevention, especially in cases of high-risk people such as smokers and former smokers, as a key to success in overcoming the disease. Early diagnosis leads to better treatment, and prevention speaks for itself.

Several research groups are trying to develop a simple blood test that would detect lung cancer cells early before they spread. One approach is to use mass spectrometry to identify proteins found only in certain cancers. Cancer cells have a number of specific proteins on their surface that are not found in healthy cells. These proteins suffer immune system attacks via a number of antibodies that attach themselves to the proteins. So if someone develops lung cancer, certain antibodies should be present in their blood. Being able to identify these antibodies has been difficult, but researchers have identified some surface proteins specific to small cell lung cancer and are in the process of improving techniques to identify more types of cancer-related protein antibodies.

Another group of researchers are developing an early-identification technique that involves washing cancer cells from the lungs with salt water. Researchers found that when they rinsed out part of the lung containing the tumor with salt water during surgery, the salt-water rinse contained many of the cancerous genetic mutations found in the tumors. The key to this approach is to be able to identify the genetic mutations that occur in cancer development and then test for each one. If it ultimately works, the lung-washing procedure theoretically could be done on an out-patient basis to screen patients before they develop lung cancer tumors and need surgery.

Scientists have also identified a gene mutation that may help predict the odds of surviving lung cancer for five years or more. They found that non–small cell lung cancer patients with a mutation in the p53 gene (see Chapter 6), which has already been linked to cancer, were more than twice as likely to die of the disease within four years as patients without the mutation. They also found that the mutation occurred more frequently in patients who both smoke and drink alcohol. Learning more about the genetics of lung cancer will lead to improved treatments through gene therapy (see Chapter 6) and other approaches involving the genetic makeup of cancerous and healthy cells. For example, in the case of the p53 gene, if doctors can test for and find its presence, they may be more aggressive in treating patients because they have a higher statistical chance of dying.

Prevention efforts for lung cancer go beyond anti-smoking campaigns and early detection to stop tumor growth. Researchers are also looking at the possibility of creating new drug regimens that would prevent lung cancer from ever developing in high-risk people. For example, early studies with a drug used to treat dry mouth, called *anethole dithiolethione* (ADT), indicate that it may reduce the risk of lung cancer in former and current smokers. The researchers have hypothesized that the drug may work like an antioxidant, as it seeks out free radicals that cause cancer and destroys them. In preliminary studies, researchers have also found that smokers and former smokers who regularly took pain-relieving drugs, specifically NSAIDs such as aspirin and ibuprofen, were less likely to develop lung cancer. The effect occurred in smokers and former smokers who took NSAIDs at least three times a week for a year or more.

SUMMARY

Advances in the fight against lung cancer are occurring yearly. Some are technical, such as the surgery thoroscopy, which requires only a 1.8 inches (4.5 centimeters) wide incision through which surgeons can remove an entire lobe of the lung. Others stem from advances in biotechnology to identify the disease's molecular foundations. Because of the now accepted recognition that lung cancer often results from genetic mutations involving several different genes, future treatment approaches will include those based on identifying and manipulating the genes that are responsible for cancer in the individual patient (see Chapter 6). A single cure or "magic bullet" for lung cancer is unlikely because of the disease's molecular complexity, but scientists have overcome higher odds. Until that day comes, better treatments in terms of efficacy, safety, and reduced side effects are on the horizon.

Smoking and Air Pollution

Breathing is the foundation of human life, but what we breathe in greatly impacts our lungs and health. Incredibly sturdy and efficient in many ways, the respiratory system also has an abundance of fragile tissue that is easily damaged by smoking tobacco and pollutants in the air. This damage can lead to an increased risk of asthma and allergies, chronic bronchitis, lung cancer, and many other respiratory diseases.

Even though our lungs and respiratory system are assaulted by a multitude of pollutants that our ancient ancestors never encountered, exposure to harmful pollutants is not a phenomenon of modern times. Forest fires, volcanos, sulfur lakes, and many other natural occurrences have contaminated the air and assaulted the lungs since the first air-breathing animals walked the earth. Human beings began to inadvertently harm their lungs when they began using fire for cooking and warmth.

Despite being the smartest animals on Earth, people knowingly risk disease and death by breathing in smoke from cigarettes, cigars, and pipes. Smoking causes the majority of lung cancer, emphysema, and chronic bronchitis cases. Not only does smoking increase the risk of lung disease but also the risk of illnesses such as heart disease, stroke, and oral cancer.

Diseases caused by smoking kill more than 440,000 people in the United States each year. Furthermore, secondhand smoke exhaled by smokers and emitted from burning cigarettes accounts for 3,000 lung cancer deaths each year in people who do not smoke. So why do people keep smoking? Because it becomes an addiction.

A person can choose not to smoke, but breathing is mandatory. According to tracking data by the U.S. Environmental Protection Agency (EPA), five of the six principal air pollutants—carbon monoxide, lead, particulate matter, sulfur dioxide, and volatile organic compounds—have all decreased significantly since the passage of the 1970 Clean Air Act. The only major pollutant that hasn't decreased is nitrogen oxide. Nevertheless, by the 1990s, American industry was still emitting more than 2.4 billion pounds of toxic pollutants into the atmosphere.

According to the CDC an estimated 50,000 to 120,000 premature deaths each year are associated with air pollution exposure. Air pollution appears to affect asthma sufferers greatly, especially children with asthma. The CDC has also stated that air pollution leads to more than 100 million days of restricted activity for the total number of people with asthma in the United States. New research is further defining the effects of air pollution on health, especially respiratory health. For example, in 2002 researchers for the first time linked long-term exposure to fine particulate matter from coal-fired power plants, factories, and diesel trucks to an increased risk of dying from lung cancer.

This final chapter takes a closer look at many aspects of smoking and air pollution. It discusses how both of these respiratory assailants affect populations and looks at some of the specific health affects attributed to them. This chapter includes information on both smoking and secondhand smoke and on what scientists have learned about the addictive aspects of smoking. Air pollution is discussed in terms of its history, constituents, and growing body of scientific evidence concerning damaging health effects. Although air pollution is mostly associated with outside air pollutants, this chapter also looks at indoor air pollution, which still greatly affects the health of many people, especially those in third-world countries.

SMOKING

Smoking is recognized as the single most preventable cause of disease, disability, and death in the United States. In addition to its well-known relationship with lung cancer and COPD, including bronchitis and emphysema, cigarette smoking is also associated with cardiovascular disease, stroke, and cancers of the mouth, pharynx, larynx, esophagus, pancreas, uterine cervix, kidney, and bladder. The cause of one-third of all cancer deaths in the United States, smoking has also been linked to a wide variety of other conditions and disorders, including infertility, impotence among men, slow wound healing, and peptic ulcer disease.

In addition to causing these diseases in smokers, secondhand smoke causes lung cancer and other health problems in nonsmokers and is classified by the U.S. Environmental Protection Agency as a known human (Group A) carcinogen. Smoking in pregnancy also adversely affects many fetuses and newborns. It accounts for an estimated 20 to 30 percent of low-birth weight babies, up to 14 percent of preterm deliveries, and some 10 percent of all infant deaths. Even babies who are born apparently healthy often have narrowed airways and reduced lung function because their mothers smoked during pregnancy.

Parents who smoke can also adversely affect the health of their children after they are born. Scientists have documented that children with parents who smoke have exacerbation of asthma, increased frequency of colds and ear infections, and a higher rate of sudden infant death syndrome (SIDS). According to the American Lung Association, secondhand smoke from parents and other relatives causes 150,000 to 300,000 cases of lower respiratory tract infections in children less than 18 months of age. These infections lead to up to 15,000 annual hospitalizations of children each year.

The approximately 440,000-plus deaths each year in the United States due to smoking represent one in every five American deaths (see Figure 8.1). According to the CDC, every year in the United States premature deaths rob more than six million years from the collective potential lifespan of all those who die from smoking that year. Furthermore, the American Lung Association estimates that smoking costs the United States approximately $150 billion each year in health care costs and lost productivity.

Scientists have discovered that cigarette smoke contains more than 4,000 chemicals, with approximately sixty of them labeled as "carcinogens," or

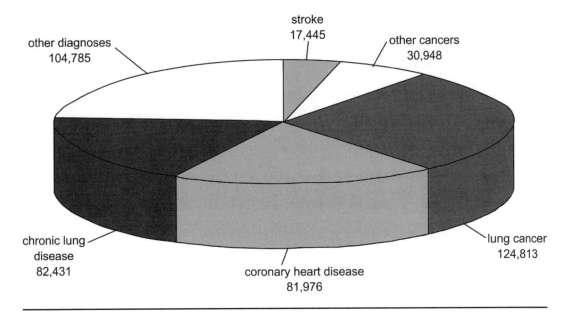

Figure 8.1. Deaths due to smoking.
Annual numbers of U.S. deaths from diseases attributable to cigarette smoking for the years 1995–1999. Data from Centers for Disease Control and Prevention.

cancer-causing chemicals. Among the many chemicals found in cigarette smoke are ammonia, arsenic, carbon monoxide, DDT, formaldehyde, nitric acid, volatile hydrocarbons, and alcohols.

Cigarette smoking also creates tar in the lungs. "Tar" is the term used to describe the particulate matter that is generated by burning tobacco and the many other chemical components of cigarettes. As a component of cigarette smoke, tar is a bituminous product made up primarily of nitrogen, oxygen, hydrogen, carbon dioxide, carbon monoxide, and a range of volatile and semi-volatile organic chemicals (substances that can emit fumes or vapors at room temperature), including polycyclic hydrocarbons such as benzopyrene, hydrocyanic acid, nitric oxide, creosols, and amphenols. In its condensed form, it is an oily, dark-colored, sticky substance that is related heavily to lung and throat cancers in smokers.

Cigarette companies and advertisers have implied that low-tar, low-nicotine cigarettes are safer and have fewer ill effects on health. But scientific research has shown that the use of low-nicotine and low-tar cigarettes often leads smokers to heavier smoking and deeper inhaling to make up for the reduced tar and nicotine content, which may result in the smoker actually taking in more tar and nicotine. Researchers have also found that a switch to low-tar, low-nicotine cigarettes does not decrease the smoker's risk of lung cancer, emphysema, heart attack, or other diseases.

Despite these facts, more than 46 million Americans, approximately 23.3 percent of the population, still smoke. Most people start smoking socially, perhaps as a result of peer pressure when they were young or because of advertising and other marketing that told them smoking was "cool" (see "Did You Know? Teen Smoking"). Although more than thirty million current smokers have said they want to quit smoking, the vast majority of them do not. The primary reason is that nicotine is an extremely addictive drug. When inhaled in smoke, it reaches the brain even faster than drugs given intravenously. Furthermore, smokers also associate smoking with many so-

Did You Know? Teen Smoking

The younger people are when they start smoking, the greater their risk of developing lung cancer. One study found that smoking as a teenager can cause permanent genetic changes in the lungs and forever increase the risk of lung cancer, even if the smoker subsequently stops.

cial activities, including lighting up after a meal and smoking during parties and other social activities.

A Short History

Modern tobacco plants are believed to have first appeared in the Americas around 6000 BCE. Historians speculate that by 1 BCE Native Americans had started to find uses for tobacco, including smoking it. One of the most widely recognized human social habits in the world today, smoking in Europe and the rest of the civilized Western world began with the discovery of America in the fifteenth century. Christopher Columbus observed the native Indians smoking tobacco through a pipe and rolled into cigars (see illustration). Before long, tobacco use was introduced to Europeans, and pipe smoking became a popular pastime with the British aristocracy.

The cultivation of tobacco became the economic basis of the first successful English colonies in North America. By 1630, tobacco was the Virginia farmer's primary crop. The French am-

Tobacco label showing a Native American presenting Christopher Columbus with a sheaf of tobacco leaves, c. 1866. Courtesy of the Library of Congress.

bassador to Portugal, Jean Nicot, took a liking to smoking and extolled it as having medicinal values, including curing gout and even cancer, among many others ailments. In fact, tobacco became known as "nicotiana" in Nicot's honor. (The ingredient "nicotine" was also named after him.) By the 1570s, many physicians were convinced that tobacco had numerous beneficial effects on health.

Tobacco eventually became popular among the lower classes as well and spread throughout the world. However, not everyone was enamored with tobacco and smoking. England's King James I pronounced in 1604 in his "Counterblast to Tobacco" that smoking was "loathsome" to view, "hateful" to smell, and "dangerous to the lungs." Although it was more than 100 years before the U.S. surgeon general would declare smoking harmful to a person's health, in 1836 Samuel Green wrote in the *New England Almanack and Farmer's Friend* that tobacco smoke was a "poison." Green also noted, "A man will die of an infusion of tobacco as of a shot through the head."

The number of people smoking cigarettes in the United States grew rapidly after the Civil War when a type of tobacco called "Bright" tobacco from a uniquely cured yellow leaf was used to make a milder and more aromatic tobacco that was easy to grow. By the late 1880s, the first practical cigarette-making machine for the individual smoker had also been introduced, which further increased tobacco use. Nevertheless, the popularity of cigarette smoking is really a phenomenon of the twentieth century. Highly effective advertising campaigns and new means of mass producing cigarettes, thus reducing their cost, led to cigarette smoking skyrocketing in the early 1900s.

A Growing Concern

Almost parallel with the growth in cigarette smoking, more and more reports were published concerning the hazards of smoking. German researchers in 1930 made the first statistical correlation between cigarette smoking and cancer. In 1938, researchers at Johns Hopkins University found evidence that smokers did not live as long as nonsmokers. The American Cancer Society followed with an announcement in 1944 that smoking was likely linked to lung cancer, although it had no strong scientific proof at the time.

In the early 1950s, however, the *Journal of the American Medical Association* stated that smoking was definitely related to lung cancer and COPD. The general public first became widely aware of the dangers of smoking in 1952, when *Reader's Digest* printed an article entitled "Cancer by the Carton." Soon other publications followed up on this news, and in 1953 cigarette sales declined for the first time in more than two decades.

In 1964, the first U.S. surgeon general's report on the effects of cigarette smoking on health appeared. In 1966, the surgeon general required that all packs of cigarettes include a label warning that smoking could adversely affect a person's health, including warnings about lung cancer, heart disease, emphysema, and complicated pregnancy.

Cigarette sales began to decline after the surgeon general's report, and the tobacco industry counteracted by creating low-tar and low-nicotine cigarettes. However, in 1971, cigarette advertising on television was banned and tobacco consumption once again dropped, this time by 10 percent.

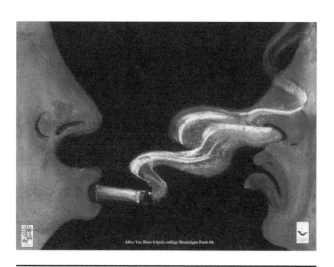

A French anti-smoking poster showing secondhand smoke. © National Library of Medicine.

Cigarette advertising in general that appeals to children, such as the former "Joe Camel" advertising campaign, has also been banned.

Growing public awareness about the health risks of tobacco use has led to a continued decline in smoking among men. However, the number of women in America who smoke has increased over the years, and cigar smoking has become popular again.

The Physiological Effects

Although smoking affects many parts of the body, it has its most damaging effects on the respiratory system and especially the lungs. Smoking causes many physiological changes in the airways and lungs that lead to diseases and impaired respiratory functioning. As smoke passes through the bronchi, hydrogen cyanide and other chemicals that it contains cause an inflammation in the lining of the bronchi.

According to a study in Great Britain, smoke-induced inflammation in the respiratory system may result from increased neutrophils, the primary blood leukocyte that performs phagocytosis, that is, consumes and destroys foreign materials. According to the study, smoking a single cigarette results in an acute increase in neutrophils at one hour. Although normally these cells serve to help protect the respiratory system, their overabundance results in the bronchi becoming narrower and reduced airflow. This process leads to a weakening of the bronchi's functioning and a higher risk of bronchial infections.

In addition to affecting the many cells that produce mucus in the lungs and bronchi, causing them to grow in size and number, tobacco smoke also contains a sticky tar-like substance. This tar covers the cilia in the respiratory system, which serve to help keep the system clean of foreign particles. Tar slows down the cilia's "sweeping" action, which helps to get rid of harmful substances that we breathe in, such as dirt and bacteria. Smoke also contains chemicals that kill off cilia cells, thus reducing the number of cilia available to help clean the airways and lungs. The increase in mucus and the decrease in cilia cells work in tandem to help clog the airways with excess mucus. In addition to causing smokers cough, this excess mucus can easily become infected.

The long-term effects of smoking include destroying many tissues in the lungs. For example, this tissue destruction reduces the number of alveoli and blood vessels in the lungs, which leads to less oxygen being carried throughout the body.

Many chemicals are also absorbed into the blood through the lungs. As a result, the carbon monoxide in the smoke attaches to hemoglobin, which is the molecule in the blood that helps carry oxygen throughout the body. This replacement of some oxygen in the blood further hinders the gas exchange process so important to overall good health. For example, physical activity such as exercise, sports, and even walking or working in the garden requires oxygen-rich blood, and physical endurance tests have shown that smokers reach exhaustion much earlier than nonsmokers.

The link between smoking and lung cancer is well known. Cigarette smoke contains many carcinogens, or cancer-causing chemicals, that stick to the lung in the form of tar. The carcinogens in the tar can damage the cells and cause them to turn into cancerous growths over a period of time. However, scientists are still trying to determine exactly how these changes occur and why some smokers develop lung cancer while others do not (see "Did You Know? Gender Differences in Disease Risk from Smoking").

In the case of lung cancer, researchers have found that the chemical benzopyrene in cigarette smoke directly attacks and damages the p53 gene (see Chapter 6). This gene helps to block cancer and prevent cancer cells from growing. Studies have also shown that the nicotine in tobacco may affect the synthesis of the **neurotransmitter acetylcholine** by lung cancer. In a study of patients with small cell lung cancer, researchers found that some lung tumor cells readily respond to acetylcholine and that the receptors in these cells are activated by nicotine. Acetylcholine may act as a growth factor fuelling the continuous growth and division of cancerous cells. As a result, nicotine might make lung cancers more aggressive by stimulating tumor cells to grow and divide, thus increasing the chances of lung cancer spreading.

Smoking's effects on the immune system can also contribute to lung cancer and many other diseases, from respiratory infections to cancers in other parts of the body. Cigarette tar, for example, contains benzene derivatives that can block DNA synthesis and inhibit the synthesis of various immune system cells, such as T cells, which are needed to fight off infections, cancers, and other diseases. Components of tar are also extremely potent inhibitors of various **cytokines**, a class of immunoregulatory substances, blocking their production by the immune system. As a result, some scientists believe that cigarette smoke blocks both cytokine production and lymphocyte proliferation, thus acting like immunosuppressive agents used in organ transplantation. Smoking may also produce free radicals that can damage

Did You Know? Gender Differences in Disease Risk from Smoking

Men who smoke increase their risk of death from lung cancer by more than twenty-two times and from bronchitis and emphysema by nearly ten times. Women who smoke increase their risk of death from lung cancer by nearly twelve times and from bronchitis and emphysema by more than ten times.

immune system cells and DNA. Free radicals are by-products of oxygen metabolism and attack DNA, proteins, and fats. Under normal circumstances, the healthy body uses an assortment of antioxidant vitamins, minerals, and enzymes to control free radicals.

While this book focuses on the respiratory system, it would be remiss not to mention the many other ill effects of smoking. The effects of smoking on the heart are devastating. Nicotine raises blood pressure and also aids in the formation of blood clots. In addition to its effect of depleting the blood's oxygen supply, carbon monoxide in smoke can lead to the development of cholesterol deposits on artery walls. The cumulative effects of these occurrences add up to an increased risk of heart attack and poor circulation, which can lead to a stroke, loss of circulation in fingers and toes, and impotence in men. Smoking also causes increased stomach acid secretion, leading to heartburn and ulcers. The body eliminates carcinogens found in smoke via urine. As a result, smoking can cause bladder cancer, which is often fatal. The high blood pressure associated with smoking can also damage the kidneys.

Overall, scientists believe that smoking affects the entire body in a variety of ways. For example, studies using PET have found that smoking can cause a 33 to 46 percent reduction in monoamine oxidase B (MAO-B) levels in the heart, lungs, kidneys, and spleen. Monoamine oxidase B is an enzyme involved in breaking down neurotransmitters in the brain and plays a role in several disease processes, particularly age-related neurological disorders. It is also a target for many antidepressant drugs and drugs that treat neurological disorders. Although researchers do not yet know the significance of such decreased levels in smoking, the study clearly demonstrates that smoking impacts the entire body and not just the respiratory system.

Environmental Tobacco Smoke

Environmental tobacco smoke (ETS), more popularly known as passive, involuntary, or secondhand smoking, occurs when nonsmokers breathe in other people's tobacco smoke. Environmental tobacco smoke includes both the smoke exhaled by the smoker and the smoke that escapes from the lit tobacco. Studies have shown that nonsmokers exposed to ETS absorb nicotine and other harmful compounds just as smokers do.

Environmental tobacco smoke has been linked to a wide variety of illnesses and health effects that were once thought only to affect smokers themselves. The first scientific evidence associating ETS with lung cancer in nonsmokers and respiratory problems in children came in 1986, with landmark reports published by the U.S. surgeon general and the Expert Committee on Passive Smoking of the National Academy of Sciences' National Research Council. In 1993, the EPA followed up with a major assessment of the risks of passive smoking on respiratory health. The results were based on the review of more than thirty epidemiological studies that looked specifically at passive smoking as well as information on active smoking. A wide range of other data were also analyzed, including data from animal studies.

In addition to its conclusion that 3,000 nonsmokers die from lung cancer annually in the United States, the EPA found a wide range of adverse health effects due to ETS exposure in children, who are more vulnerable to ETS because of their smaller airways. They include:

- increased risk of lower respiratory tract infections, such as bronchitis and pneumonia, that result in an estimated 150,000 to 300,000 cases annually, with 7,500 to 15,000 requiring hospitalization

- higher prevalence of fluid in the middle ear, a sign of chronic middle ear disease

- irritation of upper respiratory tract associated with a small but significant reduction in lung function

- increased frequency of episodes and severity of symptoms in asthmatic children

- higher risk for children developing asthma

Since the EPA's report, numerous studies have been conducted to confirm and expand upon the findings. For example, in 1998 the WHO conducted a seven-year study on the effects of ETS on lung cancer risk in European populations. The findings confirmed that passive smoking causes lung cancer in nonsmokers, with data showing an estimated 16 percent increased risk of lung cancer among nonsmoking spouses of smokers and a 17 percent increased risk for workers who were exposed to ETS in the workplace. Another major study conducted by the California Environmental Protection Agency also found numerous health effects associated with ETS exposure, including a relationship with nasal sinus cancer and SIDS. Studies in the United Kingdom and elsewhere have supported the EPA and WHO findings.

Researchers have also associated ETS with other problems, including heart disease. According to the NCI, passive smoking causes between 35,000 and 62,000 coronary heart disease deaths each year in the United States. A study published in the *British Medical Journal* found that nonsmokers exposed to ETS regularly have an 82 percent greater chance of suffering from stroke, and a 2001 study published in the *Journal of the American Medical Association* found that nonsmokers exposed to ETS for only 30 minutes experience hardening of the arteries.

In addition to the WHO findings mentioned earlier, researchers have also found evidence that nonsmoking employees exposed to ETS at the workplace are at greater risk of getting smoking-related health problems. For example, the CDC conducted a study that showed employees who were exposed to secondhand smoke on the job are 34 percent more likely to get lung cancer. Employees routinely exposed to large amounts of secondhand smoke, such as bartenders and waiters, can have three times as high a risk of developing lung cancer than the general population.

Other research findings from various studies include:

- Young children have nearly a 75 percent increased risk of developing respiratory illnesses if they live in households where both parents smoke.

- Women who don't smoke increase their risk of getting lung cancer by 25 percent if they live with a smoker.

- Tobacco specific carcinogens appear in the blood and urine of nonsmokers exposed to ETS.

- Pregnant women who smoke are at greater risk for low-birthweight infants, fetal problems, infant deformities, and miscarriages.

Despite the evidence, some scientists still debate the dangers of passive smoking. For example, in 2003, the *British Medical Journal* published a study that indicated nonsmokers who lived with smokers did not experience a significantly increased risk of cancer or heart disease. The study has come under intense criticism concerning its scientific analyses. Many scientists who study ETS have discounted the study, arguing that the evidence that passive smoking is harmful to nonsmokers is overwhelming.

The Addiction

In 1989, the U.S. surgeon general published a report that stated nicotine in cigarettes and other tobacco products, such as cigars, pipes, and chewing tobacco, is addictive. In fact, inhaling nicotine via tobacco smoke is considered to be as addictive as taking heroin or cocaine. For example, studies have revealed that animals cannot discriminate between the effects of nicotine and the effects of cocaine. Considering the 46 million-plus people who smoke in the United States, nicotine is the second most used and abused substance in America. Only caffeine consumed in coffee, soft drinks, and tea is a more widely used drug than nicotine (see "Did You Know? Nicotine").

Did You Know? Nicotine

Nicotine is such a powerful narcotic that one drop of purified nicotine on a person's tongue will result in death. Nicotine is so lethal that it has been used as a pesticide for centuries.

When inhaled in smoke, nicotine reaches the brain within ten seconds, making its effects as immediate as those experienced by intravenous drug users. This quick trip to the brain occurs when nicotine enters the lungs, where it is transferred to the blood and circulatory system. Nicotine is also absorbed through the skin and mucosal lining of the mouth and nose.

After reaching the brain, it can act both as a stimulant and a sedative so that people who crave these effects are attracted to it. In animal studies, researchers have found that this dual-effect phenomenon may result from how much a person smokes. It may be that lower doses affect certain neurotransmitters that help produce a stimulating effect. On the other hand, higher doses may overwhelm this initial response by producing a calming opioid-like effect, which takes longer to occur but also lasts longer.

Many smokers become addicted because smoking appears to have the ability to decrease their stress and anxiety levels. When a person is stressed, scientists have found that the hypothalamus in the brain attempts to counteract the stress by releasing increased levels of the b-endorphin protein, which has an analgesic (pain-killing) effect, and the adrenocorticotropic hormone (ACTH). Nicotine has been found to cause increased levels of ACTH as well as cortisol (a derivative of cortisone), which scientists believe to have stress- and anxiety-relieving qualities.

Overall, nicotine activates parts of the brain's circuitry that regulate feelings of pleasure, often called the "reward pathways" of the brain. For example, a nicotine molecule is shaped like acetylcholine, a neurotransmitter involved in helping the body perform many functions. Acetylcholine also plays a role in releasing other neurotransmitters and hormones that affect mood. Nicotine attaches to acetylcholine receptors in the brain and mimics the neurotransmitter's actions.

Part of nicotine's pleasurable effects are also achieved because nicotine stimulates the adrenal glands, resulting in the discharge of epinephrine (adrenaline). The burst of adrenaline, in turn, stimulates the body and causes the release of glucose, as well as an increase in blood pressure, respiration, and heart rate.

Scientists continue to delve into the numerous physiological factors, especially the neurological changes, that cause people to become addicted to nicotine and its pleasurable effects. For example, researchers have found that the hormone corticosterone acts to reduce the effects of nicotine, resulting in the need to consume more nicotine to get the desired effects.

Recent research has also shown that nicotine increases the level of the neurotransmitter dopamine, which affects the brain's circuitry involving feelings of pleasure. Ongoing research is also beginning to focus on other psychoactive ingredients in tobacco that may help make it addictive. Some studies have found that smoking may decrease the levels of monoamine oxidase (MAO), an enzyme that helps break down dopamine. Because of lower MAO levels, higher levels of dopamine result, which may further cause smokers to crave tobacco to sustain high dopamine levels. Experiments have revealed that nicotine has very little effect on MAO, indicating that there must be another chemical in tobacco or cigarette smoke that is causing this effect.

Researchers have also found that some people are more likely than others to become addicted to different substances. In the case of nicotine, they have identified a factor that may make some people less susceptible to addiction. Studies have shown that people with a genetic variant that decreases the function of an enzyme called CYP2A6, which metabolizes nicotine and makes it available in the blood for a longer period of time, are less likely to smoke or succumb to nicotine's addictive effects. This discovery has led scientists to study the possibility of using certain drugs, such as methoxsalen (currently used for skin disorders), to reduce the enzyme's activity and make it easier for people to quit smoking.

The addictive qualities of nicotine and perhaps other components of tobacco smoke are also exacerbated by the way nicotine is used by smokers. As pointed out earlier, nicotine gets to the brain rapidly. According to studies, the typical smoker takes ten puffs over the usual five minutes that it takes to smoke the cigarette. If a person smokes a pack (twenty cigarettes) a day, they are getting 200 nicotine "hits" to the brain each day. Scientists believe this is a strong factor in contributing to smoking's addictive nature.

Finally, nicotine is associated with social habits that may trigger the desire to smoke. In other words, people become accustomed to having a cigarette while drinking, going to a party, after meals, while having a conversation, and so on. In effect, the physical act of smoking becomes a type of comforting ritual in itself. These various "triggers" make it even more difficult to quit smoking.

According to statistics by the National Institute on Drug Abuse (NIDA), thirty-five million smokers try to quit each year, but less than 7 percent quit for more than a year; in fact, most relapse within a few days of quitting (see "Smoking Stats at a Glance"). The primary reason for relapse is the various withdrawal symptoms that people experience when they try to quit. These symptoms run the gamut from psychological and emotional symptoms to physical symptoms, including irritability, restlessness, difficulty concentrating, aggressiveness, depression, and increased hunger.

Many approaches and programs are available to quit smoking. Some research indicates that quitting should be a gradual process in which the smoker begins to cut back on the number of cigarettes smoked per day. This approach lessens the severity of withdrawal symptoms. Nicotine chewing gum, nicotine transdermal patches, and a drug (Zyban) are also available to help lessen withdrawal symptoms from nicotine. Some studies have shown that high abstinence rates are achieved with combined pharmacological-psychological treatment.

Smoking Stats at a Glance

- Statistics indicate that the average five minutes it takes to smoke a cigarette shortens the smokers' life by seven to eleven minutes.

- An estimated 4.5 million teenagers in the United States smoke (see Figure 8.2).

- If a child reaches age 18 without becoming a smoker, his or her odds of remaining smoke-free are around 90 percent.

- Although smoking rates have fallen among men they have increased in women, and approximately 22.2 million American women are smokers.

- More women die annually from lung cancer than any other type of cancer, with lung cancer accounting for an estimated 65,700 female deaths in 2002, compared with 39,600 estimated female deaths caused by breast cancer.

- The prevalence of smoking is highest in Native Americans/Alaskan Natives, with 40.89 percent of that population being smokers; 24.3 percent of African Americans and whites are smokers, and 18.1 percent of Hispanics smoke.

- It is estimated that 40 percent of men who are heavy smokers will die before they reach retirement age as compared to only 18 percent of nonsmokers.

- Compared with nonsmokers, cigar smokers who inhale are eight times more likely to develop oral cancer, four to ten times more likely to develop esophageal cancer, and ten times more likely to develop laryngeal cancer.

- Smoking five cigars a day is associated with the same lung cancer risk as smoking a pack of cigarettes a day.

- Fewer than one in ten people who try to quit smoking succeed.

Scientists have also developed an experimental nicotine vaccine that may help prevent nicotine from reaching the brain and, as a result, reduce its addictive effects. The vaccine consists of a nicotine derivative attached to a large protein. Experiments with rats demonstrated that the amount of nicotine that reached the brain was reduced by 64 percent.

The Good News

Despite the many health consequences associated with smoking and the difficulty that smokers have in quitting, the good news is that quitting even after years of smoking is beneficial. Furthermore, the positive effects of quitting start almost immediately. For example, within twenty minutes of hav-

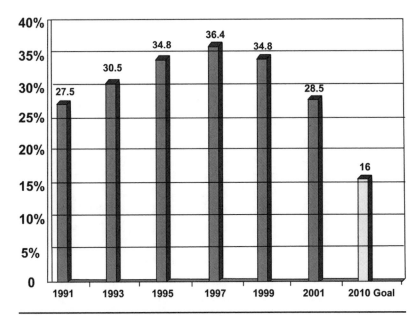

Figure 8.2. High school smokers.
Percentage of U.S. high school students who reported current cigarette smoking for the years 1991–2001. Data from Centers for Disease Control and Prevention.

ing that last cigarette, a smoker's blood pressure drops. Within eight hours, carbon monoxide levels in the blood drop to normal and oxygen levels increase to normal. Studies also indicate that the chances of heart attack decrease within twenty-four hours of quitting. Other almost immediate benefits include an improved sense of smell and taste and the noticeable ability to breathe easier.

The longer-term benefits of quitting smoking are even more important in terms of adding healthy productive years to life. For example, the excess risk of coronary heart disease associated with smoking decreases by 50 percent after one year of not smoking. Stroke risk is reduced to that of a nonsmoker in five to fifteen years after quitting. Within five years, the risk of cancer of the mouth, throat, and esophagus is half that of a smoker.

Studies have also shown that lung cancer death rates fall to about half that of people who continue to smoke. Furthermore, people who quit smoking for a long period of time significantly reduce their risk of death from chronic bronchitis, emphysema, and types of cancer.

AIR POLLUTION

Although primarily associated with modern industrial societies, air pollution has been affecting the human respiratory system for thousands of years. For example, upon examining ancient Egyptian mummies, scientists found signs of anthracosis. Also known as miner's lung, anthracosis is the accumulation of carbon and silica from inhaled smoke or dust in the lungs. In the case of the Egyptians, the exposure probably came from indoor smoke emanating from fires used for cooking and warmth.

Although we normally think of air pollution as the result of human activities, it also occurs naturally. Phenomena such as dust storms, earthquakes, and volcanic eruptions can produce enormous amounts of dust and gases in the air. Nevertheless, pollution resulting from human endeavors has been the major source of concern associated with human health.

Historically, a significant increase in air pollution occurred in the thirteenth century, when a growing population in London, England, and other cities throughout Europe led to the use of coal instead of wood as a primary energy source. Soon the urban skies were filled with smoke. With the Industrial Revolution of the eighteenth and nineteenth centuries, even more coal was burned to supply power, and urban levels of pollution reached extremely high levels. When certain weather conditions set in, pollution escalated further to create urban smog. Soon, doctors and other health care professionals began to correlate an increase in death and illnesses with increased air pollution.

Throughout the first half of the twentieth century, pollution problems grew with the burning of fossil fuels for heating and electric power generation and by automobile and other combustion engines. As a result, many urban areas soon were suffering from major air pollution problems. In Pittsburgh, Pennsylvania, for example, pollution from the city's steel mills often caused the sky to darken by noon, and office workers were known to take an extra white shirt to work with them so they could change after their shirts became blackened by soot.

For the most part, however, few formal scientific studies focused on the effects of air pollution on respiratory and overall human health. People generally expressed little concern over air pollution unless they already suffered from respiratory ailments, in which case the pollution noticeably affected their health.

There were reminders, however, of just how dangerous air

Smoke issues from the stacks of industrial buildings along a river. Courtesy of the Library of Congress.

pollution could be. In 1948 in Donora, Pennsylvania, a combination fog and temperature inversion (air temperatures increase with altitude instead of decrease) caused sulfur dioxide fumes from steel mills to become trapped over the town. The result was thousands of temporary illnesses and twenty deaths due to respiratory or cardiovascular problems. Perhaps the most stunning occurrence took place in London in 1952. An anticyclone (region of high atmospheric pressure) settled over the city creating a thick fog that lasted for five days. Called the Great London Smog, it resulted in 4,000 "extra" deaths than usual. The deaths were attributed to a seven-fold increase in sulfur dioxide and a three-fold increase in smoke. Researchers found that the peak in deaths coincided with the peak in both smoke and sulfur dioxide pollution levels. Once again, the deaths were associated with respiratory and cardiovascular problems due to pollution.

The first air pollution control agency in the United States was formed in Los Angeles County, California, in 1947. The agency began its efforts to control air pollution with such efforts as banning backyard trash incinerators and regulating industries, such as requiring oil refiners to reduce sulfur emissions and recycle wastes. In 1955, the federal Air Pollution Control Act was enacted. The act authorized the Secretary of Health, Education, and Welfare to work toward a better understanding of the causes and effects of air pollution.

In 1970, the United States government passed the Clean Air Act with the intent of improving air quality in the United States. The act established various guidelines and goals for industrial sources, power plants, and automobile factories to decrease the amount of air pollution and emissions. The Clean Air Act was amended in 1977 and 1990 and includes standards against common pollutants, including carbon monoxide, sulfur dioxide, nitrogen dioxide, lead, particulate soot, and **ozone**.

Since the passage of the Clean Air Act, air quality in the United States has improved. For example, emissions of toxic lead have dropped 98 percent, sulfur dioxide emissions by 35 percent, and carbon monoxide by 32 percent. Nevertheless, air pollution still poses a potential health problem for many people living in the United States. According to the American Lung Association's *State of the Air 2003* report, 137 million Americans live in counties with unhealthy levels of air pollution, including nearly 2 million children who suffer from asthma attacks and an estimated 1.5 million people with emphysema.

A Pollution Primer

Air pollutants are usually placed within two categories. Primary pollutants, like sulfur dioxide and carbon monoxide, are emitted directly into the atmosphere. Secondary pollutants are formed by the chemical interactions between two or more primary pollutants or between primary pollutants and a normal atmospheric element. Ozone pollution, for example, occurs when primary pollutants from factories and cars mix in the sunlight and undergo a photochemical reaction.

The Clean Air Act focused on establishing **ambient air** standards for seven "criteria pollutants." These are:

- Sulfur dioxide—Although released by natural processes such as volcanoes, sulfur dioxide also enters the air through the burning of fossils fuels, such as coal and oil. In high concentrations, their effects include irritation of the respiratory system, increased mucus production, cough, and shortness of breath.

- Nitrogen oxide—This pollutant is formed in high-temperature combustion processes where fuels are burned using air as an oxidant and is mostly associated with automobile exhaust. Nitrogen oxides have been related to irritation of the pulmonary tract and decreased lung functioning.

- Carbon monoxide—Produced when fossil fuels are incompletely burned, carbon monoxide occurs naturally and is also a byproduct of automobile exhaust and various industries; it also helps to generate ozone. Carbon monoxide can reduce the ability of the blood to carry oxygen and is extremely toxic in high quantities.

- Lead—Along with mercury, lead is one of the most dangerous metallic air pollutants and comes largely from paints and fuel combustion. Since the government banned the use of lead in paints and fuels, much lower levels reach the environment. Lead is a cumulative poison known to affect the central nervous system.

- Particulates—Particulate matter is the term for particles found in the air, including dust, dirt, soot, smoke, and liquid droplets. Particulate matter can come from fuel and biomass (plant material and animal waste used as fuel) combustion stemming from industries and automobile exhaust, as well as construction, mining, agricultural, and natural processes. Research indicates that particulate matter may have harmful effects on the respiratory system.

- Photochemical oxidants—This type of pollutant includes ozone as a secondary pollutant. An especially reactive form of oxygen, ozone is normal in the upper atmosphere but causes damage to plant and animal tissues when present at lower levels in the atmosphere.

- Hydrocarbons—Also associated with automobile exhaust, hydrocarbons have been linked with lung cancer and mucous membrane irritation.

Indoor Air Pollution

Discussions about air pollution often focus on the air we breathe outside, but indoor air pollution, even in our own homes, also presents a problem. According to the WHO, a general rule of thumb is that a pollutant released indoors is 1,000 times more likely to reach a person's lungs than outdoor pollutants. The EPA also has noted that the air inside the home can be five times more polluted than the air outside a home and can trigger asthma.

For years, indoor air pollution was primarily a concern in developing countries where wood and other biomass fuels (cow dung, crop residues, and grass) are still used in cooking and heating. But indoor air pollution also became a growing problem in modernized countries, especially the United States, when buildings were made to be increasingly airtight to reduce heat gain or loss in response to the energy crisis of the 1970s.

Many modern household products, such as carpeting and plywood, contain volatile substances. These substances are irritating or toxic and can build up to toxic levels in insulated, air-tight, and poorly ventilated buildings, which prevent them from escaping. In addition, indoor air pollutants have been associated with cooking, cleaning, and other household chores. Other sources of indoor air pollution that can affect the respiratory system include tobacco smoke, dust and dirt, mold and mildew, aerosol products, and household chemical cleaners.

The natural product of radioactive decay of uranium, called radon, is also a concern. This decay occurs in small amounts in almost all rock and soil, as well as in bricks and concrete made from these materials. Radon can seep into basements and enclosed spaces below ground level. The problem varies widely depending on the region of the country, the climate, and the building itself. Radon has been linked to an increased risk for lung cancer, and the EPA has recommended radon testing of homes.

The Research

The issues surrounding air pollution and its health effects, including its effects on the respiratory system, are highly charged. According to some, pollution is an extremely complex mixture, and not enough scientific data is available to single out any one component as being responsible for certain health effects. On the other hand, other scientists assert that statistical analysis can decipher the specific health effects of individual pollutants, even though pollutants always occur in combinations.

Further complicating the issue of air pollution is the fact that many pollutants are the result of industry and lifestyle, such as everyone driving to work in their own car instead of carpooling. As a result, discussing air pollution includes such issues as economic impact and freedom of choice. One of the issues under debate is whether or not air pollution levels are still at high enough levels to cause harm.

Few would argue, however, that the components cited in air pollution can cause serious health effects if they occur in high enough concentrations. Pollutants can cause lung injury and inflammatory reactions in the airways, leading to such respiratory problems as bronchitis and asthma. Air pollutants absorbed into nasal passages tissues can also cause allergic rhinosinusitis. Scientists also widely recognize that airborne pollutants exacerbate airway disease in persons with asthma, and that children and seniors seem to be at particular risk from air pollution.

Despite the debate, numerous studies are revealing more and more that air pollution in the United States can still reach levels that harm human health. Researchers are also beginning to correlate specific components of air pollution with certain respiratory and other health problems.

One area of growing concern is the health effects of particulate matter, even though it has been greatly reduced in many areas of the country because of environmental regulations. Nevertheless, scientific studies have linked fine particulate matter in the atmosphere to aggravated asthma, increased respiratory symptoms like coughing and difficult or painful breathing, chronic bronchitis, decreased lung function, and even premature death. Both the EPA and the WHO have expressed concern over particulate matter levels in the atmosphere. According to the Natural Resources Defense Council, approximately 64,000 people may die prematurely from cardiopulmonary causes linked to particulate air pollution each year. Again, the issue has been hotly debated because many industries produce particulate matter pollution, and further controls are seen as damaging to the economy. Nevertheless, in 2003 the Health Effects Institute released a report confirming the findings of many studies showing that increased particulate air pollution can increase daily deaths and hospitalizations.

As concern over air pollution and the many issues around it grow, numerous studies have begun to focus on air pollution and health. Many have shown a direct link between respiratory health and air pollution, including:

- A study in the February 1999 issue of the *American Journal of Respiratory and Critical Care Medicine* showed that living or working for many years in areas with high levels of particulate matter air pollution is associated with a 28 percent higher risk of dying from any type of nonmalignant respiratory disease.

- Published in the March 2002 issue of the *Journal of the American Medical Association,* results from a sixteen-year study showed a strong link between long-term exposure to fine particles of air pollution from coal-fired power plants, factories, and diesel trucks, with an increased risk of dying from lung cancer.

- In the October 2000 issue of the *American Journal of Respiratory and Critical Care Medicine*, the results of a ten-year study that monitored levels of major pollutants in Southern California since 1993 found that many common air pollutants slow children's lung development over time.

- A 2002 EPA report titled *National Air Toxics Assessments* concluded that one in 100,000 Americans will contract cancer as the result of air pollution.

- In 2002, University of Southern California scientists announced that they had found the first evidence suggesting that ground-level ozone is a "causative factor" in the development of childhood asthma.

- In the March 12, 2002, issue of *Circulation: A Journal of the American Heart Association,* researchers for the first time showed that air pollution negatively affects blood vessels. In the study, twenty-five healthy people inhaled elevated concentrations of fine particles plus ozone for two hours. After exposure, volunteers' blood vessels constricted between 2 percent and 4 percent on average. Their vessels did not constrict when they were exposed to ozone-free and particle-free air.

The air today is certainly a lot cleaner than it was in the 1950s, when factories indiscriminately belched toxic smoke and gas, and other fuels contained many more components such as lead that were harmful to health. For most healthy people, the levels of air pollution seldom have any long-term effects on their health. However, people with respiratory and heart problems are at increased risk, especially on days when pollution levels reach higher levels in many of the country's major metropolitan areas.

Despite overall improved air quality, many questions still exist concerning air pollution and public health. For example, while short-term effects of air pollution on healthy people may be negligible, scientists are still studying the effects on healthy people who experience long-term exposure to moderate air pollution levels. More scientific data is also needed on how air pollution affects specific populations, including people with respiratory diseases, the elderly, and children.

Acronyms

AAFP	American Academy of Family Physicians	**CF**	cystic fibrosis
AAT	alpha-1 antitrypsin	**CFTR**	cystic fibrosis trans-membrane conductance regulator
ACTH	adrenocorticotropic hormone	**CHART**	continuous hyperfractionated accelerated radiotherapy
ADP	adenosine diphosphate		
AMA	American Medical Association	**CoA**	acetyl coenzyme A
		COPD	chronic obstructive pulmonary disease
AMP	adenosine monophosphate	**COX-1**	cyclooxygenase 1
ARDS	acute respiratory distress syndrome	**COX-2**	cyclooxygenase 2
		CPAP	continuous positive airway pressure
ATP	adenosine triphosphate		
ATRA	all-trans retinoic acid	**CPR**	cardiopulmonary resuscitation
BCG	Bacille Calmette-Guerin	**CT**	computerized tomography
CDC	Centers for Disease Control and Prevention	**DNA**	deoxyribonucleic acid

ECLAP	Early Lung Cancer Action Program	**MRI**	magnetic resonance imaging
ECMO	extracorporeal membrane oxygenator	**mRNA**	messenger RNA
EGFR	epidermal growth factor receptor	**NAD**	nicotinamide adenine dinucleotide
EPA	Environmental Protection Agency	**NADH**	reduced nicotinamide adenine dinucleotide
ETS	environmental tobacco smoke	**NCCAM**	National Center for Complementary and Alternative Medicine
FAD	flavin adenine dinucleotide	**NCI**	National Cancer Institute
FAP	familial adenomatous polyposis	**NHLBI**	National Heart, Lung and Blood Institute
FDA	Food and Drug Administration	**NIDA**	National Institute on Drug Abuse
GSTM1	glutathione S-transferase M1	**NIH**	National Institutes of Health
HAPE	high-altitude pulmonary edema	**NMDA**	N-methyl-D-aspartate
ICAM	intracellular adhesion molecule	**NSAID**	non-steroidal anti-inflammatory drug
ILD	interstitial lung disease	**PAH**	pulmonary arterial hypertension
IMO	intravenous membrane oxygenator	**PDT**	photodynamic therapy
IMPV	intermittent negative pressure ventilation	**PET**	positron emission tomography
MAO	monoamine oxidase	**RDA**	recommended dietary allowance
MDR-TB	multidrug resistant tuberculosis	**RNA**	ribonucleic acid

SARS	severe acute respiratory syndrome	**SLC**	secondary lymphoid tissue chemokine
SCLC	small cell lung cancer	**TB**	tuberculosis
SIDS	sudden infant death syndrome	**WHO**	World Health Organization

Glossary

Acetylcholine A neurotransmitter that is a derivative of choline and is released at the ends of nerve fibers in the somatic and parasympathetic nervous systems.

Adenosine triphosphate (ATP) A nucleotide derived from adenosine that occurs in tissue. The "molecular currency" of intracellular energy transfers, ATP stores and transports chemical energy within a cell and is a precursor for RNA formation.

Adenovirus A group of viruses that can cause various respiratory diseases such as pharyngitis and influenza. However, adenoviruses can be genetically altered so that they do not cause infections but do target specific proteins, making them a potentially effective vector for carrying genes to certain parts of the body.

Aerobic Anything having to do with acquiring oxygen from the air. In terms of exercise, for example, aerobic refers to increased oxygen consumption by the body to enhance respiratory efficiency.

Allergies Hypersensitive reaction to a particular substance or allergen; symptoms vary in intensity.

Alveoli Tiny porous air sacs found in human lung tissue formed at the end of bronchioles.

Ambient air An environmental term meaning any unconfined portion of the atmosphere, as in open air or surrounding air.

Angiogenesis inhibitor A substance that may prevent the formation of blood vessels, such as preventing the growth of blood vessels from surrounding tissue to a solid cancerous tumor. Preventing blood vessel formation helps keep cancer from spreading.

Antibodies Proteins used by the immune system to identify and counteract harmful foreign substances such as viruses and bacteria; antibodies are normally produced in response to an antigen (macromolecule) that is unique to their target.

Antihistamine Any substance that works by counteracting the effects of histamine on a receptor site, including medicines used to treat allergies, hypersensitive reactions, and colds.

Aorta The largest artery in the human body. It supplies oxygenated blood from the left ventricle of the heart to the branching arteries, which in turn supply oxygen to all parts of the body.

Aortic bodies Chemoreceptors found in the aortic arch, the curved portion between the ascending and descending parts of the aorta.

Apoptosis Cell death that occurs when the cell uses a specialized cellular machinery to kill itself. Apoptosis occurs naturally throughout biology as a way for organisms to control cell numbers and eliminate cells that threaten survival.

Arteries Tubular structures that transport blood away from the heart to capillaries throughout the body's various systems and tissues. Arteries deliver blood that has passed through pulmonary circulation in the lungs and has become saturated with oxygen.

Asphyxia A lack of oxygen and/or excess of carbon dioxide in the body. Asphyxia is most often caused by interruption of breathing and unconsciousness.

Autonomic nervous system Part of the nervous system that helps control involuntary functions and actions, usually involving the smooth muscles, heart, and glands. The autonomic nervous system helps control many variations in respiration.

Biomarkers Identifiable biochemical indicators of a biological process or event.

Bohr effect High concentrations of carbon dioxide and hydrogen ions in the capillaries in metabolically active tissue decrease the affinity of hemoglobin for oxygen. Leads to a shift to the right in the oxygen dissociation curve.

Bronchi The two large air tubes leading from the trachea to the lungs that convey air to and from the lungs.

Bronchiole Any of the smallest bronchial tubes that end in alveoli.

Bronchoscope A slender, flexible, tubular instrument used to view the bronchi, or pass other instruments into them.

Capillaries The smallest blood vessels that form a network throughout the body and its tissues. The micro-thin walls of capillaries easily allow oxygen to pass into cells and for cell waste products to pass into the blood from cells for eventual elimination via expiration.

Carbamino compounds Blood proteins that bind to carbon dioxide.

Carbocyclic Characterized by a ring composed of carbon atoms.

Carcinogens Cancer causing agents.

Carcinoids Small tumors, usually found in the gastrointestinal tract, that secrete serotonin.

Carotid arteries Two major arteries branching from the aorta that pass on either side of the neck and supply oxygenated blood to the brain.

Carotid bodies Chemoreceptors found in the carotid arteries.

Catheter A thin flexible tube inserted into the body to introduce or withdraw fluids or to keep passageways open.

Cauterization Destroying tissue with heat, cold, or caustic substances. Cauterization is usually used to seal off blood vessels or ducts.

Chemoreceptors Cells that respond to changes in their chemical environment by creating nerve impulses. Some chemoreceptors in the brain respond to carbon dioxide levels in the blood to help regulate breathing.

Chemotherapy To use chemical agents to treat or control disease.

Chest percussion A technique that aids clearance of secretions up and out of the lungs and increases the amount of air entering the lungs. A cupped hand is used to clap the chest firmly.

Chloride shift Describes the exchange of negatively charged chloride ions for negatively charged bicarbonate ions across an erythrocyte's cell membrane.

Cilia Hairlike projections from the surface of a cell. In the respiratory system, cilia help filter out foreign particles from the air before they reach the lungs.

Coenzyme A small molecule that is not a protein but may be a vitamin that is essential for the activity of some enzymes.

Computerized tomography (CT) A method of creating cross-sectional images of the human body by taking data from multiple x-ray images and compiling them into pictures through the use of a computer. CT scanning can reveal soft-tissue and other structures not viewable with conventional x-rays.

Conchae Structures or parts that resemble a sea shell in shape. In the respiratory system, conchae are three bony ridges or projections—the superior, middle, and inferior conchae—on the surface of the nasal cavity sides.

Cystic Fibrosis A condition distinguished by the development of excess fibrous connective tissue in an organ. For example, certain inflammatory diseases can cause fibrosis in the lungs.

Cytochromes Class of membrane-bound intracellular hemoprotein respiratory pigments. These enzymes function in electron transport as carriers of electrons.

Cytokines Any of various proteins secreted by cells of the immune system that serve to regulate the immune system.

Cytoplasm The contents of cells inside the cell membrane but outside of the nuclear membrane. The cytoplasm includes the internal fluid of the cell (cytosol) and a variety of organelles.

Cytotoxic agents Agents that are toxic to cells.

Defibrillator An electronic mechanism that administers an electric shock of preset voltage to the heart through the chest wall in an attempt to restore the normal rhythm of the heart.

Electron transport system A complex sequence found in the mitochondrial membrane that accepts electrons from electron donors (such as the NADH energy molecule) and then passes these electrons across the mitochondrial membrane creating an electrical and chemical gradient. This gradient provides energy for ATP synthesis.

Enzyme A complex protein produced by cells that acts as a catalyst in specific biochemical reactions.

Epidemic A widespread outbreak of an infectious disease that affects a disproportionately large number of people within a given population.

Epidemiology The study of the demographics of disease processes that are common enough for the application of statistical tools, including epidemics and other diseases.

Epiglottitis Inflammation of the epiglottis (the flap of cartilage that covers the windpipe while swallowing) characterized by fever, severe sore throat, and difficulty swallowing.

Erythrocytes Red blood cells.

Flavoproteins The enzymes that contain flavin bound to a protein. Flavoproteins play a major role in biological oxidations.

Free radical A chemically active atom or group of atoms that contain a chemical charge due to an excess or deficient number of electrons. In the human body, free radicals are most often oxygen molecules that have lost an electron and will stabilize themselves by stealing an electron from a nearby molecule. Free radicals are likely to take part in strong chemical reactions and can damage large molecules within cells.

Gas exchange In the respiratory system, gas exchange refers to the process of acquiring oxygen from the air and eliminating carbon dioxide from the blood.

Genomics The branch of genetics that studies organisms in terms of their genomes (full DNA sequences).

Glycolysis A metabolic process connected with respiration that breaks down carbohydrates and sugars in cells through a series of reactions to either pyruvic acid or lactic acid, and releases energy for the body in the form of adenosine triphosphate (ATP), the universal energy currency of the cell.

Goblet cell An epithelial cell that secretes mucus.

Haldane effect High concentration of oxygen, such as occurs in the alveolar capillaries of the lungs, promotes the dissociation of carbon dioxide and hydrogen ions from hemoglobin. Leads to a shift to the left in the oxygen dissociation curve.

Hemes The deep red organic pigment that contains iron and other atoms to which oxygen binds in blood hemoglobin. Hemes are found in most oxygen-carrying proteins.

Hemoglobin Red respiratory protein of red blood cells (erythrocytes). It functions in oxygen transport to tissues and helps in transporting carbon dioxide from tissues back to the cells for elimination during expiration.

Hilus A depression or fissure where vessels or nerves or ducts enter a bodily organ.

Histamine Amine (a compound derived from ammon) formed from histidine (an essential amino acid found in proteins that is important for the growth and repair of tissue) that stimulates gastric secretions and dilates blood vessels. Histamines are released by the human immune system during allergic reactions.

Hyperventilation An increased and excessive depth and rate of breathing greater than demanded by the body's needs; can lead to abnormal loss of carbon dioxide from the blood, dizziness, tingling of the fingers and toes, and chest pain.

Immune system A body system that includes the thymus and bone marrow and lymphoid tissues. The immune system protects the body from foreign substances and pathogenic organisms in the form of specialized cellular responses.

Immunosuppressive drugs Drugs that suppress the immune system and immune responses. Immunosuppressive drugs are often used in transplantation to keep the patient's immune system from rejecting transplanted organs and tissue.

Immunotherapy A treatment based upon the concept of triggering the body's own natural defenses to fight off the disease, usually by stimulating the immune system.

Inflammatory immune response A response by the immune system that occurs when tissues are injured by bacteria, trauma, toxins, heat, or any other cause. Chemicals including histamine, bradykinin, serotonin, and others are released by damaged tissue. These chemicals cause blood vessels to leak fluid into the tissues, resulting in localized swelling. This helps isolate the foreign substance from further contact with body tissues.

Inflammatory mediators Soluble, diffusible molecules that act locally at the site of tissue damage and infection.

Intercostal muscles Muscles found under the ribs that play a role in respiration. The external intercostals are found between the ribs with fibers running down. They pull ribs together and raise the rib cage during inspiration. The internal intercostal muscles are also between the ribs; their fibers run at right angles to the external fibers. The internal intercostals depress the rib cage during forced expiration.

Ion An atom or group of atoms that carries a positive or negative electric charge as a result of having gained or lost one or more electrons.

Kinins Various polypeptide hormones, such as bradykinin, that are formed locally in the tissues and cause dilation of blood vessels and contraction of smooth muscle.

Krebs cycle A series of enzymatic reactions in cell mitochondria, which involve oxidative metabolism of acetyl compounds to produce high-energy phosphate compounds for storage in phosphate bonds (as in ATP), and are the source of cellular energy.

Laryngotracheal groove A furrow at the caudal end (tail or hind part) of the floor of the embryonic pharynx that develops into the respiratory tract.

Laryngotracheobronchitis Inflammation of the larynx, trachea, and bronchial passageways.

Left atrium One of the four chambers of the human heart, which also include the left ventricle, right ventricle, and right atrium. The left ventricle is the left upper chamber of the heart that receives oxygen-rich blood from the lungs and pulmonary veins and then passes the blood to the left ventricle, which sends the oxygen-rich blood through the aorta to all parts of the body except the lungs.

Left ventricle One of the four chambers of the human heart, which also include the right ventricle, left atrium, and right atrium. As the main pumping chamber of the heart, the left ventricle is on the left side of heart below the left atrium and receives oxygen-rich blood from the left atrium and pumps it throughout the body, except to the lungs.

Leukocyte A white blood cell that is part of the immune system and can destroy bacteria and fungi.

Leukotrienes Biologically active compounds that are part of a body's immune reactions in response to physiological conditions such as inflammation.

Live virus vaccine A vaccine containing a "living" virus that is able to give and produce immunity, usually without causing illness.

Lymph glands The small bean-shaped masses of lymphoid tissue that are surrounded by a capsule of connective tissue, are distributed along the lymphatic vessels, and contain numerous lymphocytes, which filter the flow of lymph. Also called *lymph nodes.*

Lymph vessels Vascular ducts that carry lymph, which is eventually added to the venous blood circulation.

Lymphatic system The interconnected system of spaces and thin tubes that branch, like blood vessels, into tissues throughout the body. Lymphatic vessels carry the colorless fluid called lymph, which helps transport infection-fighting lymphocytes. It is part of the body's immune system and helps protect the lungs and other parts of the respiratory system from infections and disease.

Macrophage A type of white blood cell, or leukocyte, that is part of the immune system and primarily acts to help clear the blood of bacteria and other particles it recognizes as foreign invaders.

Mass spectrometry An instrumental method for identifying the chemical constitution of a substance by means of the separation of matter (gaseous ions) according to their atomic and molecular mass.

Messenger RNA (mRNA) The form of RNA that carries information from DNA in the nucleus to the ribosome sites of protein synthesis in the cell. mRNA is the key intermediary in gene expression, translating the DNA's genetic code into the amino acids that make up proteins.

Metastasize To spread through the body, as in cancer cells.

Millimeter(s) of mercury A unit of pressure equal to that exerted by a column of mercury at 0°C one millimeter high at mean sea level. It equals 1/760 (0.001316) atmosphere.

Mitochondria Organelles containing enzymes that have the main function of converting the potential energy of food molecules into adenosine triphosphate (ATP), the universal energy currency of the cell. Each mitochondrion is composed of folds called *cristae*, which give a much increased surface area on which chemical reactions can occur.

Mucociliary Pertaining to mucus and to the cilia of the epithelial cells in the respiratory system.

Neurochemical An organic substance or chemical found in the nervous system that performs any variety of activities, such as transmitting nerve impulses among neurons.

Neurons Nerve cells. These primary cells of the nervous system are located in the brain, the spinal cord, and in the peripheral nerves.

Neurotransmitter A molecule that is used to transmit signals between nerve cells or neurons. Neurotransmitters play a vital role in sending signals to ensure that the body properly functions.

Nucleic acid Any of various acids, such as deoxyribonucleic acid (DNA) and ribonucleic acid (RNA), that are composed of nucleotide chains and are the vital components of all living things.

Oxaloacetic acid An acid formed by oxidation of maleic acid, as in metabolism of fats and carbohydrates in the Krebs cycle.

Oxidation-reduction reaction A reaction in which there is transfer of electrons from an electron donor (the reducing agent) to an electron acceptor (the oxidizing agent). Also called the redox reaction. In the electron transport system, this reaction results in molecules alternately losing and gaining an electron.

Oxidize Add oxygen to or combine with oxygen, usually in chemical processes.

Oxygen dissociation curve A graph that shows the percent saturation of hemoglobin at various partial pressures of oxygen. The curve shifts to the right (the Bohr effect) when less than a normal amount of oxygen is taken up by the blood and shifts to the left (the Haldane effect) when more than a normal amount is taken up. Factors influencing the graphic curve include changes in the blood pH and the partial pressure of carbon dioxide. The curve can also be affected by temperature, carbon monoxide, and certain disease states.

Oxyhemoglobin Oxygenated hemoglobin found in arterial blood. It is the oxyhemoglobin that releases oxygen into tissues.

Ozone An extremely reactive form of oxygen that forms naturally in the atmosphere via a photochemical reaction. It acts as an air pollutant in the lower atmosphere but is a beneficial part of the upper atmosphere.

Pandemic An epidemic that occurs over a large geographic area, sometimes throughout the world.

Paramyxovirus Any virus of the family that cause measles, mumps, and the respiratory syncytial virus (RSV), which causes pneumonia and the parainfluenza virus.

Parasympathetic nervous system Part of the autonomic nervous system, mostly associated with the involuntary actions to induce secretion, increase the tone and contractility of smooth muscle, and to slow heart rate.

Partial pressure The specific pressure exerted by individual components of a gas mixture, commonly expressed in millimeters of mercury (mmHg).

Pathogens Disease-producing agents such as virus, bacterium, or other microorganisms.

pH Measure of acidity and alkalinity of a solution. pH is represented as a number on a scale on which a value of 7 represents a neutral level, lower numbers indicate increased acidity, higher numbers increased alkalinity. The normal pH range of blood is 7.35–7.45. Variations as small as a few tenths of a pH unit from this range can cause serious illness and even death. The respiratory system helps balance pH levels.

Photochemical reaction Any reaction produced by the action of light.

Placebo A pill or treatment that contains no active medication. Placebos are often prescribed to ease the patient's anxiety as opposed to any actual effect, and they are also used in research as control substances to compare the effectiveness of other drugs.

Plasma The colorless watery fluid portion of blood in which particulate components are suspended, including red blood cells (erythrocytes), white blood cells (leukocytes), and platelets.

Pleura A membrane that envelopes the lung and attaches the lung to the thorax. There are two pleurae, right and left, entirely distinct from each other, and each pleura is made of two layers. The parietal pleura lines the chest cage walls and covers the upper surface of the diaphragm, and the visceral pleura tightly covers the exterior of the lungs. The two layers are actually one continuous sheet of tissue that lines the chest wall and doubles back to cover the lungs. The pleura is moistened with a thin, serous secretion which helps the lungs to expand and contract in the chest.

Porphyrin A complex, nitrogen-containing compound that makes up the various pigments found in living tissues. Iron-containing porphyrins are called *hemes*.

Prophylactic Anything designed to prevent or slow the course of an illness, disease, or condition.

Prostaglandin Powerful substance similar to hormones and produced in response to trauma. These extremely potent mediators of diverse physiological processes can affect blood pressure and metabolism and smooth muscle activity.

Protein Any of a group of nitrogenous organic compounds that are essential constituents of living cells and are essential to numerous essential biological compounds, including enzymes and hormones. In the respiratory system, for example, hemoglobin is an oxygen carrying protein.

Pulmonary Relating to or affecting the lungs.

Pulmonary artery Artery that transports oxygen-poor blood from the heart's right ventricle to the lungs.

Pulmonary function tests Tests to evaluate the mechanical properties of the lung by studying lung volumes and capacities, such as total lung capacity (TLC), residual volume (RV), tidal volume (Vt), and more.

Right atrium One of the four chambers of the human heart. The right atrium receives oxygen-depleted blood from the body and passes it to the right ventricle.

Right ventricle One of the four chambers of the human heart. The right ventricle takes oxygen-depleted blood from the right atrium and pumps it into the lungs to undergo the gas exchange process of gaining oxygen and eliminating carbon dioxide.

Septum A partition, dividing wall, or membrane that separates bodily spaces or masses of tissue. In the respiratory system, septum most often refers to the cartilage separating the two nostrils.

Surfactant A substance that acts on the surface of objects. In the respiratory system, surfactants are secreted by type-II pneumocyte cells into the alveoli and respiratory air passages. This surfactant helps make pulmonary tissue elastic in nature. In the case of alveoli, it helps allow them to expand and contract without damage.

Systemic capillaries The network of capillaries in the body tissues that allows exchange of gases, nutrients, and wastes between the bloodstream and body. Blood transports oxygen to the systemic capillaries, where oxygen moves through the thin capillary walls into the cells. Carbon dioxide passes through the capillary walls from the cells.

Throat culture A technique for identifying disease bacteria in material taken from the throat. A throat culture is obtained by wiping the throat with a cotton swab.

Tracheobronchial Relating to the tracheal and bronchial tubes

Vaccine A preparation of killed microorganisms, living-attenuated organisms, or living-fully virulent organisms to initiate or artificially increase immunity to a particular disease.

Vagus nerve The tenth of twelve cranial nerves. It originates somewhere in the medulla oblongata in the brainstem and extends down to the abdomen. Perhaps the most important nerve in the body, the vagus nerve tells the brain to "turn off" its signals for inspiration when the lungs are fully expanded.

Veins Tubular branching vessels that carry blood from the capillaries towards the heart.

Wegener's granulomatosis A rare disease of unknown cause that is characterized by granuloma formation in the respiratory tract, glomerulonephritis, and necrotizing granulomatous vasculitis. Wegener's granulomatosis is often associated with interstitial lung disease.

X-ray diffraction The scattering of x-rays by the atoms of a crystal, creating a diffraction pattern that shows the structure of the crystal. X-ray diffraction is one method of studying proteins in cells.

Organizations and Web Sites

Alliance for Lung Cancer Advocacy, Support, and Education (ALCASE)
500 W. 8th Street, Suite 240
Vancouver, WA 98660
Phone: (800) 298-2436
http://www.alcase.org

American Academy of Allergy, Asthma and Immunology
611 East Wells Street
Milwaukee, WI 53202
Phone: (414) 272-6071
Patient Information and Physician Referral Line: (800) 822-2762
http://www.aaaai.org/

American Academy of Family Physicians
P.O. Box 11210
Shawnee Mission, KS 66207-1210
Phone: (800) 274-2237
http://www.aafp.org/

American Association for Respiratory Care
11030 Ables Lane
Dallas, TX 75229
Phone: (972) 243-2272
Fax: (972) 484-2720, (972) 484-6010
Email: info@aarc.org
http://www.aarc.org/

American Lung Association
61 Broadway, 6th Floor
New York, NY 10006

Phone: (212) 315-8700
http://www.lungusa.org/

American Medical Association
515 N. State Street
Chicago, IL 60610
Phone: (312) 464-5000
http://www.ama-assn.org/cgi-bin/feedtool.pl

Asthma and Allergy Foundation of America (AAFA)
1233 20th Street NW
Suite 402
Washington, DC 20036
Phone: (202) 466-7643, or (800) 7-ASTHMA
Fax: (202) 466-8940
Email: info@aafa.org
http://www.aafa.org/

National Cancer Institute
NCI Public Inquiries Office
Suite 3036A
6116 Executive Boulevard, MSC8322
Bethesda, MD 20892-8322
Phone: (800) 4-CANCER (800-422-6237)
http://www.nci.nih.gov/

National Emphysema Foundation
c/o Sreedhar Nair, M.D.
15 Stevens Street
Norwalk, CT 06850
Phone: (203) 854-9191
Email: gary@emphysemafoundation.org
http://www.emphysemafoundation.org/

National Heart, Lung and Blood Institute (NHLBI)
NHLBI Health Information Center
P.O. Box 30105
Bethesda, MD 20824-0105
Phone: (301) 592-8573
Email: nhlbiinfo@rover.nhlbi.nih.gov
http://www.nhlbi.nih.gov/

National Institute on Drug Abuse
6001 Executive Boulevard, Room 5213
Bethesda, MD 20892-9561
Phone: (301) 443-1124
Email: Information@lists.nida.nih.gov
http://www.nida.nih.gov/

National Institutes of Health (NIH)
9000 Rockville Pike
Bethesda, MD 20892
Phone: (301) 496-4000
Email: NIHInfo@od.nih.gov
http://www.nih.gov/

National Lung Health Education Program
HealthONE Center
899 Logan Street
Suite 203
Denver, CO 80203-3154
Email: nlhep@aol.com
http://www.nlhep.org/

ADDITIONAL WEB SITES

Lung Cancer Online
http://www.lungcanceronline.org

A gateway to lung cancer resources for the benefit of people with lung cancer and their families that is intended to facilitate the time-consuming and often frustrating process of learning about lung cancer, treatment options, and support services.

Lungcancer.org
http://www.lungcancer.org

Provides lung cancer information online.

National Library of Medicine
PubMed
http://www.ncbi.nlm.nih.gov/entrez/query.fcgi

NHLBI Asthma Management Model System
http://www.nhlbisupport.com/asthma/index.html

An information management tool designed to facilitate science-based decision-making and evidence-based medicine in long-term asthma management.

Office of the Surgeon General's Tobacco Cessation Guideline
http://www.surgeongeneral.gov/tobacco/

The latest drugs and counseling techniques for treating tobacco use and dependence.

Pulmonary Channel
http://www.pulmonarychannel.com/

Contains information on various pulmonary conditions.

Pulmonary Education Research Foundation
http://www.perf2ndwind.org/

Education, research, and information resource for people with chronic respiratory diseases.

Virtual Hospital
http://www.vh.org/
Includes detailed information on the respiratory system and respiratory disease.

Bibliography

Abbey, D. E., et al. "Long-term Inhalable Particles and Other Air Pollutants Related to Mortality in Nonsmokers." *American Journal of Respiratory and Critical Care Medicine* 159, no. 2 (February 1999): 373–382.

"American Lung Association State of the Air 2003 Report." http://lungaction.org/reports/sota03_full.html.

Belongia, E. A., R. Berg, and K. Liu. "A Randomized Trial of Zinc Nasal Spray for the Treatment of Upper Respiratory Illness in Adults." *American Journal of Medicine* 111, no. 2 (August 2001): 103–108.

Bernardi, L., et al. "Effect of Rosary Prayer and Yoga Mantras on Autonomic Cardiovascular Rhythms: Comparative Study." *British Medical Journal* 323, no. 7327 (December 22–29, 2001): 1446–1449.

Bourke, S. J. *Respiratory Medicine*, 6th ed. Malden, MA: Blackwell Pub., 2003.

Brigham, Kenneth L. *Gene Therapy for Diseases of the Lung*, vol. 104, New York: Marcel Dekker, 1997.

Brook, R. D., et al. "Inhalation of Fine Particulate Air Pollution and Ozone Causes Acute Arterial Vasoconstriction in Healthy Adults." *Circulation* 105, no. 13 (April 2, 2002): 1534–1536.

Cellular Respiration (interactive multimedia). Paradise, CA: CyberEd, Inc., 1996–1999.

Enstrom, J. E., and G. C. Kabat. "Environmental Tobacco Smoke and Tobacco Related Mortality in a Prospective Study of Californians, 1960–98." *British Medical Journal* 326, no. 7398 (May 17, 2003): 1057.

Gauderman, W. J., et al. "Association Between Air Pollution and Lung Function Growth in Southern California Children." *American Journal of Respiratory and Critical Care Medicine* 162, no. 4 (October 2000): 1383–1390.

Gilgoff, Irene S. *The Breath of Life: The Role of the Ventilator in Managing Life-Threatening Illnesses*. Lanham, MD: Scarecrow Press, 2001.

Hess, Dean, and Robert M. Kacmarek. *Essentials of Mechanical Ventilation*. 2nd ed. New York: McGraw-Hill, Health Professions Division, 2002.

"History of Medicine." National Library of Medicine. http://www.nlm.nih.gov/hmd/hmd.html/.

Hlastala, Michael P., and Albert J. Berger. *Physiology of Respiration.* 2nd ed. New York: Oxford University Press, 2001.

Knight-Lozano, C. A., et al. "Cigarette Smoke Exposure and Hypercholesterolemia Increase Mitochondrial Damage in Cardiovascular Tissues." *Circulation* 105, no. 7 (February 19, 2002): 849–854.

Kozower, B. D., et al. "Immunotargeting of Catalase to the Pulmonary Endothelium Alleviates Oxidative Stress and Reduces Acute Lung Transplantation Injury." *Nature and Biotechnology* 21, no. 4 (April 2003): 392–398.

Mainous, A. G., and W. J. Hueston. "Variation in Use of Broad-Spectrum Antibiotics for Acute Respiratory Tract Infection." *Journal of the American Medical Association* 289, no. 21 (June 4, 2003): 2797; author reply 2797–2798.

Mohr, U., et al. *Relationships Between Respiratory Disease and Exposure to Air Pollution.* Washington, DC: ILSI Press, 1998.

Otsuka, R., et al. "Acute Effects of Passive Smoking on the Coronary Circulation in Healthy Young Adults." *Journal of the American Medical Association* 286, no. 4 (July 25, 2001): 436–441.

"Pollution-Caused Cell Messengers Linked to Disease," *Immunotherapy Weekly,* October 17, 2003, 3.

Pope, C. A., et al. "Lung Cancer, Cardiopulmonary Mortality, and Long-Term Exposure to Fine Particulate Air Pollution." *Journal of the American Medical Association* 287, no. 9 (March 6, 2002): 1132–1141.

"Sars Outbreak Contained Worldwide," World Health Organization. http://www.who.int/mediacentre/releases/2003/pr56/en/.

"Treating Respiratory Spray to Treat Lung Cancer." *Cancer Weekly,* November 1, 1999.

Turner, R. B., and W. E. Cetnarowski. "Effect of Treatment with Zinc Gluconate or Zinc Acetate on Experimental and Natural Colds." *Clinical and Infectious Diseases* 131, no. 5 (November 2000): 1202–1208.

Verrier, R. L., M. A. Mittleman, and P. H. Stone. "Air Pollution: An Insidious and Pervasive Component of Cardiac Risk." *Circulation* 106, no. 8 (August 20, 2002): 890–892.

Zhang, Y., et al. "Positional Cloning of a Quantitative Trait Locus on Chromosome 13q14 that Influences Immunoglobulin E Levels and Asthma." *Nature Genetics* 34, no. 2 (June 2003): 181–186.

Index

About the Author

DAVID PETECHUK is an independent scholar. He has written on numerous topics including genetics and cloning and has worked with scientists and faculty in areas such as transplantation and psychiatry.